D1352790

THE WAR OF THE SEXES

THE WAR OF THE SEXES

How Conflict and Cooperation Have Shaped Men and Women from Prehistory to the Present

PAUL SEABRIGHT

PRINCETON UNIVERSITY PRESS
Princeton and Oxford

Published by Princeton University Press, 41 William Street, Princeton, New Jersey 08540
In the United Kingdom: Princeton University Press, 6 Oxford Street, Woodstock, Oxfordshire OX20 1TW
press.princeton.edu

Jacket photograph: *Bride and groom figurines on cake top back to back.* Photographer: John Slater. Courtesy of Getty Images.

Library of Congress Cataloging-in-Publication Data

Seabright, Paul.
The war of the sexes : how conflict and cooperation have shaped men and women from prehistory to the present / Paul Seabright.
p. cm.
Includes bibliographical references and index.
ISBN 978-0-691-13301-0 (hbk. : alk. paper)
1. Sex (Psychology). 2. Sex differences (Psychology). 3. Interpersonal relations. 4. Men—Psychology. 5. Women—Psychology. I. Title.
BF692.S345 2012
306.709—dc23 2011053122

British Library Cataloging-in-Publication Data is available

This book has been composed in Minion with Knockout display by Princeton Editorial Associates Inc., Scottsdale, Arizona

Printed on acid-free paper. ∞

Printed in the United States of America

10 9 8 7 6 5 4 3 2 1

For Isabelle

CONTENTS

Acknowledgments ix

Part One Prehistory

1 Introduction 3
2 Sex and Salesmanship 27
3 Seduction and the Emotions 40
4 Social Primates 60

Part Two Today

5 Testing for Talent 93
6 What Do Women Want? 111
7 Coalitions of the Willing 126
8 The Scarcity of Charm 141
9 The Tender War 157

Notes 183
References 211
Index 233

ACKNOWLEDGMENTS

A BOOK LIKE THIS IS THE FRUIT of a great deal of teamwork. I am grateful to many people who have contributed time, energy, and expertise to sharpen the argument, unclog the prose, and remove some of the many errors in earlier drafts. The following made valuable comments on part or all of the manuscript: Terri Apter, Nicoletta Berardi, Giuseppe Bertola, Sam Bowles, Brian Boyd, Wendy Carlin, Ananish Chaudhuri, Tyler Cowen, Isabelle Daudy, Roberta Dessi, Anna Dreber-Almenberg, Jeremy Edwards, Rosalind English, John Fingleton, Guido Friebel, Muna Genikis, Ricardo Guzmán, Geoff Hawthorn, Peregrine Horden, Marie Lalanne, Maike Lentner, Eunate Mayor, Patricia Morison, Danielle Nolles, Alice Seabright, Diana Seabright, Edmond Seabright, Jack Seabright, Luke Seabright, Laura Spinney, Kim Sterelny, Charmaine Tan, Geeraj Vermeij, and Marie-Claire Villeval. The responsibility for any remaining errors is entirely my own.

Camille Boudot, Katie Brewis, Sebastian Kohls, and Adeline Lo provided outstanding research assistance. Alison Booth, Will Dawkins, Paul Hooper, and Johanna Rickne gave helpful advice and information on particular questions.

Nicoletta Berardi, Samuele Centorrino, Elodie Djemai, Guido Friebel, Astrid Hopfensitz, Marie Lalanne, and Manfred Milinski have been stimulating and tolerant coauthors of research papers reported in the book. Among my other coauthors, Wendy Carlin, John Fingleton, Geoff Hawthorn, Thierry

Magnac, Alice Mesnard, Bob Rowthorn, Mark Schaffer, and Fiona Scott-Morton have left an important mark on the way I think about the issues considered here. I have learned much on these topics from conversations (sometimes over many years) with Bina Agarwal, Sam Bowles, Rob Boyd, Geoff Brennan, Jung-Kyoo Choi, Partha Dasgupta, Jeremy Edwards, Ernst Fehr, Ben Fraser, Pauline Grosjean, Angie Hobbs, Paul Hooper, Hillard Kaplan, Willemien Kets, Jacques Le Pottier, Sheilagh Ogilvie, Ugo Pagano, Linda Partridge, Richard Portes, Carlos Rodriguez-Sickert, Suzanne Scotchmer, Claudia Senik, Kim Sterelny, and Libby Wood.

Two places have been particularly important in giving me the physical and intellectual conditions that made it possible to do this work. The Toulouse School of Economics, where I have now been based for a little over ten years (though it did not always go by that name), has been a superb environment for research, teaching, and the enjoyment of life; my debts to friends and colleagues there are too numerous to list individually. The Santa Fe Institute, at whose Behavioral Sciences Program I have been an almost-annual guest for the last seven years, is a wonderful place to meet scholars from many disciplines; every visit has been the occasion for new and sometimes quite unexpected discoveries.

Preliminary (sometimes too preliminary) versions of the ideas developed in this book were presented to audiences at the Royal Economic Society Annual Lecture in 2005; the European Economic Association's twentieth anniversary session "The Economics of the Human Animal" in 2006; the Hooker Visiting Lecture at McMaster University in 2006; the BBVA Annual Lecture 2006; the Darwin Lectures for the Darwin Anniversary in 2009 at Darwin College, Cambridge; and the Institute of Economic Growth Distinguished Lecture in New Delhi in 2011. I learned much from audience reaction on these occasions and would like to thank Bina Agarwal, William Brown, Richard Portes, John Sutton, Jaume Ventura, and Xavier Vives for the invitations. I am also grateful for the many valuable comments from seminar audiences to whom I have presented research material on these topics in Auckland, Canberra, Hong Kong, New Delhi, Norwich, Oxford, Santiago, Stockholm, and Toulouse.

My friend and agent Catherine Clarke has been, for this project as for others, an unfailing source of support and shrewd advice. My editor at Princeton, Seth Ditchik—as well as Peter Dougherty, with whom my contacts go back at least a decade—have managed to combine friendly encouragement with absolute rigor. Terri O'Prey steered the book through production with the support and professionalism of Peter Strupp and his colleagues at Princeton Editorial Associates. My thanks to them all.

In addition to those mentioned above, I am indebted in a variety of ways to Emmanuelle Auriol, Jacques Barrat, David Begg, Gloria Carnevali, Gérard Cazé, Nathalie Cazé, Sabrina Choudar, Jenny Corbett, Diane Coyle, Jacques Crémer, Christiane Fioupou, Marie-Pierre Flandy, Mireille Gealageas, Susan Greenfield, David Hart, Susan Hurley, Jacqueline Hylkema, Marc Ivaldi, Priscille Lachat-Sarrete, Jean-Jacques Laffont, Alena Ledeneva, Claudine Leduc, Jean Leduc, Jibirila Leinyuy, Michael Morgan, Damien Neven, Rory O'Connor, Nick Rawlins, Stephanie Renard, Andrew Schuller, Kay Sexton, Jean Tirole, Anne Vanhems, and Charlotte Wang for advice, encouragement, friendship, and inspiration. In small and large ways, without them the book would have been a different one.

My parents, Jack and Diana Seabright, encouraged my very first projects and continue to encourage them to this day. My children, Alice, Edmond, and Luke Seabright, have been an unfailing source of ideas as well as a formidable skeptical barrier for my own arguments to pass. Their mother, Isabelle Daudy, has never lost faith in the power of lucid discussion over domestic conflicts of interest or in the compatibility of reason and love. This book is dedicated to her.

PART ONE

PREHISTORY

ONE

Introduction

En notre vie mortelle il n'est d'autre vérité que le chevauchement, tout le restant n'étant que lanternes et fariboles.

(Our mortal life is nothing but coupling; all the rest is just lanterns and nonsense.)

—Albert Cohen, *Belle du Seigneur,* 1968

Cooperation and Conflict

A FEW MILES WEST OF CHICAGO, on a warm night in late spring, a fast and fancy courtship is playing out in full view of some admiring bystanders. He's lithe and he's loaded, and she's had her eye on him since the moment he swung into view. The admiration is clearly mutual: he's invited her to join him for a meal, with a sparkle in the eye that suggests he's looking for something in return and that he doesn't expect to receive no for an answer. Her charms are unmistakable: her voluptuous curves single her out unmissably in his eyes from the gaggle of her girlfriends fluttering excitedly about on their night out. It looks at first glance as though they understand each other perfectly, this playful couple. But in fact there's a lot they don't know about each other, things that might surprise them if they did. She doesn't realize that he's much less rich than he looks. And he doesn't realize that she's had herself cosmetically enhanced: those curves aren't as authentic as he thinks they are. He has no idea just how many of her girlfriends have done the same. And if he were capable of giving the matter a moment's thought, he might be a little put out to realize that the admiration they all share for his attributes has everything to do with his offer of dinner and nothing at all to do with his physique. He may be in it for the pleasure, but she is only too aware that it's also a business deal. She's not a gold digger exactly, but she has a shrewd head on her shoulders.

This fleeting episode of soap opera is not what it seems, either. The characters are not people but insects: specifically, dance flies of the species *Rhamphomyia longicauda*. Like human beings (as well as like some other species, including chimpanzees), these flies make a strong connection between food and sex. Male dance flies compete with one another for the attentions of females by cornering scarce food resources to offer as a sexual bribe, and they have various tricks to make their bribes look bigger than they really are. The females compete in turn for the attentions of the males with the biggest bribes by inflating their physical charms—literally. They blow their abdomens full of air to make themselves look more curvaceous, and thus more fertile, than they are. Each party offers the other something that looks better than it is, and what both receive in exchange is less impressive than they hope. Both males and females use economic strategies to strike a sexual bargain.

Human males and females do the same, though in different ways from the dance flies, in different ways from each other, and in different ways in the twenty-first century from the ways they have done in previous centuries. But they have used economic strategies for sexual purposes since the dawn of our species. By *economic strategies* I mean systematic ways of negotiating over things they value, whether these are obviously economic goods like money and food, or other, nonmonetary resources like time, effort, and self-esteem. These dance flies show us that our sexual goals enter into these economic strategies too, and that, consciously and unconsciously, we negotiate in pursuit of our sexual purposes and not just to try to enrich ourselves financially. Understanding these sexual purposes, and the opportunities and conflicts to which they lead, will help us understand better why the economic relations between men and women take the form they do. It will help us see why conflict exists and how it shapes inequalities in power between men and women, inequalities that have shifted over the millennia as economic conditions have changed. Male dance flies corner scarce food not to eat it themselves but to increase their control over the sexual choices

Dance flies (*Rhamphomyia longicauda*), male (above) and female (below). Note the rounded abdomen of the female, under the center wing. © Bruce Marlin.

of females. Human males have similarly accumulated scarce economic resources as a means of exerting control over the choices of human females. As with dance flies, the most interesting questions about economic relations between men and women are not about how much they respectively consume but about how much they each control.

Conflict exists in a particularly complicated form between men and women because human beings are the most cooperative species on earth; that's the central claim of this book. This cooperation in turn has developed because it is necessary, because the course of our evolution has increased drastically the damage we can inflict on each other if we fail to agree. Over the past few million years, the ancestors of human beings began to colonize a very risky evolutionary niche: the long childhood. It was a niche that needed a more complicated form of cooperation than anything previously attempted by any animal. Human children are the pampered movie stars of the animal kingdom: they need care for a longer period, and from a larger and more diverse team of supporting staff, than the offspring of any other species. Hollywood stars who take hours and a retinue of assistants to get dressed are prodigies of self-reliance compared to the average human newborn, who takes a year and the encouragement of many gushing relatives even to stand up. Such a complex period of dependence is an invitation to misunderstanding and conflict between the parents, as well as between the parents and the other relatives. Thus sexual encounters and their prospect of offspring are freighted with potential consequences more complex than those faced by any other animal. But unless the enterprise works, our expensively produced, pampered, and terribly vulnerable offspring will not survive, and natural selection will efface all trace of us. So sex for us, far more than for any other species, is not just about reproduction, about making new humans: it's about all the alliances and rivalries that it stimulates among the vast supporting cast of each new human who appears on the scene.

We're not the only animals who use sex for more than reproduction: among chimps and bonobos, for instance, sex plays a central role in making and breaking alliances and friendships.[1] That's why there's no puzzle about the evolution of homosexuality, which is widespread in nonhuman animals and plays a particularly powerful role in social bonding in bonobos (it seems like a puzzle only if you think the sole adaptive consequences of sexual encounters are the offspring that directly result from those encounters).[2] But we humans have built more elaborate structures of cooperation than any other animal, and sex has repercussions for all of these, so it's hardly surprising that we have also developed more elaborate strategies of deception, manipulation, and conflict.

At first glance this claim may seem wildly self-contradictory. We humans are the product of evolution by natural selection, and natural selection is the most unforgiving of designers. How can a cooperative partnership built by natural selection come to have conflicts of interest at its very core? To many observers it has seemed that sexual conflict must be a recent by-product of our civilization, about which biology has nothing to say. Yet this picture of human couples, uniquely cursed by their hyperactive brains and the confines of their artificial living conditions to be dissatisfied in their relationships while flies, lizards, birds, and bonobos copulate in untroubled sensuality, contradicts what we know about other species. Sexual conflict, far from being uniquely human, is everywhere in nature, even if in human beings it takes a very developed form.[3] Birds do it, bees do it, even educated fleas fight as they fornicate. Bedbugs, baboons, dolphins, elephant seals, spiders, scorpions, and water striders engage in rape, with the males using brute force, drugs, physical restraint, teamwork, or ingenious mechanical equipment to force themselves on reluctant females. Males of the scorpion species *Parabuthus transvaalicus* have evolved special "lite" poisons to drug their females into acquiescence, as their regular brand appears to be dangerously strong; the dance of male and female scorpions is a heady tango between danger and desire.

Females of many species deploy in turn a startling variety of counterstrategies, ranging from body armor to sperm barriers to sisterly coalition building—expensive biological investments that would be mystifying if resistance to sex were merely the product of Victorian inhibition. Male chimpanzees and dance flies provide dinner for females to bribe them into having sex, and females manipulate the gullibility of males to induce them to pay more than they want to in order to obtain less than they hope. Both males and females in outwardly monogamous species, such as fairy wrens and fat-tailed dwarf lemurs, engage in surreptitious extraconjugal sex, provoking the hypocritical and sometimes violent jealousy of their partners when they are discovered (and often when they are not).[4] Females of numerous species charm, cajole, and manipulate males into contributing to the care of their young, while males trick and cheat their way out of the explicit or implicit promises that persuaded the females to yield to their advances in the first place. Male lions and gorillas abuse and even kill the infants they believe females to have borne to other males, and the mothers' mourning has barely begun to subside before they are having apparently willing sex with their children's killers.[5] No human soap opera could outstrip in violence, hypocrisy, and manipulation the daily drama of relations between the sexes across the entire animal kingdom. Why does nature work that way, and what does it mean for us?

The strangest clue is provided by the behavior of praying mantises and many species of spider, whose females eat their males after intercourse, from the head downward, usually beginning their meal even before the male has finished ejaculating.[6] Remarkably, this is not the culmination of the sex war but its ultimate, harmonious resolution. The males usually make little effort to escape their fate, for the simple reason that they lead very solitary lives. Many die without ever meeting a single female: the lucky ones are unlikely to meet a second even if they should escape the first. Their reproductive interests therefore coincide almost totally with those of the first female they happen to meet. Natural selection is stern: in a sexually reproducing species, there's no point in trying to escape a cannibalistic female unless there's some prospect of meeting another female later. Otherwise, in an environment where food is scarce, the male's body does more service to his reproductive interests if, after ejaculating, he can offer it to his partner as a meal. It's enough to make you grateful for the invention of cigarettes.

Cannibalistic spiders are exceptional: most other animals are likely to meet more than one potential mate during a lifetime and face choices that influence how and with whom they mate. It's the element of choice that plants the seed of sexual conflict. Like a conversation at a party with someone who cannot restrain himself from looking over your shoulder to see who else there might be to talk to, sexual relations in almost all species are clouded by the possibility that either partner might be better off with someone else, now or in the future. Each partner has an interest in steering the interaction in directions that allow for those other possible encounters. With two pairs of hands at the steering wheel and two divergent itineraries, some degree of conflict is no surprise.

To say there's conflict doesn't mean that male and female interests are completely opposed—far from it. It means just that they're not completely aligned. And even a slight difference in priorities can create vast potential for mistrust. In fact, far from being antithetical to cooperation, conflict is at its most difficult and challenging precisely when cooperation has most to offer, because there's more at stake—a bigger potential pie to share and a greater temptation to hurl blame at each other if it all goes wrong. Paradoxically, therefore, sexual conflict in human beings is so intractable because we are, by nature's standards, such a spectacularly cooperative species, one whose sexual partnerships at their best achieve astonishing feats of collaboration. Implacable enmity between males and females would be a relatively easy predicament to handle compared to the mix of cooperation and conflict that we encounter: if you know that anything which benefits your opponent must harm you, it's easy to decide never to concede anything unless you strictly have to. But you and your prospective sexual partner are not opponents. You have something really important to gain from cooperating, so it's easy to be persuaded into contributing a lot to that partnership—and then to feel that you've been manipulated into doing so by someone who wanted to receive a larger share of the benefits or to contribute a smaller share of the costs.

It's worth investigating the logic of this predicament a little further. The conflicts of interest between men and women arise for two distinct reasons. The first is that when couples bargain together, or even just when they decide how much energy and effort to contribute to their shared projects, they're not completely transparent to each other. The second is that even if they were transparent, they'd be unable to commit completely to doing what they undertake to do, and that makes each of them wary about trusting the other too far. Let's look at the transparency problem first. Instead of sizing each other up, working out how the fruits of their partnership will end up being shared, and cutting straight to the resolution without any bluff or manipulation, couples face a strong temptation to engage in shadow boxing. Each of them seeks to project a mask to signal their better qualities and to protect themselves against being taken for granted by the other, and the fact that the mask may say something truthful about them doesn't make it any less a mask. But the masks can get in the way of the communication they both want. That's why sexual and amorous encounters abound with missed opportunities, regretted outbursts, and unreasonable sacrifices—choices that in retrospect seem insane. And it's worse when there's a prospect of a really important outcome: the bigger the prize, the more easily it can paralyze us.

Consider how two people who are really attracted to one another can fail to seize the opportunity or can seize it only to find it disappointing. He realizes that the day she knows she has secured him is the day her ardor will begin to cool: all that energy seems no longer quite so crucial to the outcome, and in any case she may adjust downward her view of how difficult, and therefore how desirable, a catch he is. She also knows the same will be true of him; and so, if they are lucky and their passion is evenly matched, they play a game of feigned indifference in a futile attempt to ward off the unthinkable, the end of the game. They risk missing the opportunity altogether rather than sell themselves for too low a price, and they escape disappointment only by prolonging the uncertainty for as long as they can. The nineteenth-century French novelist Stendhal's great novel *The Red and the Black* narrates a painfully prolonged version of this predicament: the two lovers are so paralyzed by the fear of revealing themselves to be less valuable catches than the other might think that they find it impossible to express love or tenderness at all. But in one way or another, any compelling soap opera has this theme at its heart. We'd never keep watching if the happy ending were either impossible or inevitable; at the same time, it's the mesmerizing attractiveness of that ending that throws so many obstacles in the way. Experimental studies have now confirmed what both novelists and soap writers already know: uncertainty about what others are feeling for us is a powerful reinforcer of sexual attraction, and effective seducers avoid appearing predictable if they possibly can.[7]

Think, too, how a marriage may founder because both parties feel that their contributions are underappreciated. So she sighs at him, and he frowns at her, just to avoid being taken for granted. They fall out of the habit of communicating delight in each other's presence, and without delight their marriage sets its course for the rocks. If the couple didn't have so much to lose, they would worry much less about signaling their respective contributions; it's their worrying so much that puts in jeopardy everything they might achieve.

It's tempting to think that transparency would solve the problem of conflict by making bluffing pointless, but even people for whom bluffing is pointless can face conflict of a second kind. This is because, however sincere their intentions, they can't commit not to change their minds. (Couples aren't alone in this: the US Congress famously cannot bind its successors.) This uncertainty inherently limits the nature of the sacrifices they are prepared to make for their relationship, even if such sacrifices are ones they'd make gladly if only they could be assured of the relationship's durability. And it also means that such sacrifices as they do make can hurt them very badly. Women may give up a career to raise children, only to find, when their marriage breaks down, that they have few marketable professional skills. Men may work long hours to earn enough to bring up their children in comfort, only to find that on divorce they lose custody of the children they had hoped to see more of once they were older. If the relationship didn't have such value while it lasted, its breakdown would not do such terrible damage, and the fear of its breakdown would not have so chilling an effect on mutual trust.

Sexual conflict, then, is the shadow cast by cooperation, and it wouldn't be so painful if we didn't have so much to share or so great a fear of being exploited in the process of sharing. Human couples fight because the human experiment in cooperation is by nature's standards so productive, so ambitious. They fight because although cooperation requires the partners to signal their needs and their talents to each other both before and after they decide to cooperate, signaling creates opportunities for manipulation and fears of being manipulated. These fears may be corrosive when they are justified and even more corrosive when they are not. Cooperation also requires a couple to hope for more lasting commitments than either is capable of making, and fears about the weakness of these commitments may be corrosive whether or not they are subsequently vindicated by events.

This book deals mostly with what's special about that human experiment and asks in particular why control of the spectacular economic resources it has made possible has been distributed so unequally between the sexes. It also asks what cooperation between the sexes has to teach us about cooperation in other contexts, such as the workplace and political life. But first we need to understand what we share with the rest of nature. Sexual conflict may be particularly intractable for us human beings, but it exists in various forms throughout the natural world.

Different Stakes for Males and Females

Sexual conflict is a fact of life for both males and females, but the two sexes react very differently to it because the stakes are different for each sex (at least in those sexually reproducing species that have two distinct and determinate sexes).[8] Put simply, males in most species have a much greater reproductive interest in the quantity of their offspring, females in their quality (with some important exceptions). This difference in priorities is the result of a simple but profound difference between male and female sex cells. Eggs are large, expensive to make, and scarce, while sperm are small, cheap, and abundant. Indeed, it's the fact of creating the larger sex cells that *defines* females as distinct from males. The contrast in abundance is dramatic: human females, for instance, release one new egg per month, while men produce around one thousand sperm per second—theoretically enough in that same month to fertilize all the women of reproductive age in the world.

The abundance of sperm means that a woman can and must be selective about its source. Her eggs, being scarce, are valuable: not only does she release only one egg per month, but if that egg is fertilized, she will bear the fetus in her body and be unable to produce any more offspring for at least a year. She will then find herself with a child whom she must feed and protect for many years; the male may credibly threaten to have nothing to do with his children, but she cannot.[9] So her opportunities to bear young are far too precious to waste on unsuitable males.

But the female's selectivity creates a challenge for the men. As well as locating a fertile woman, each man must persuade her to accept what he has to offer and compete with any other men who may be trying to do the same thing. Those who have the skills to overcome the selectivity of women and the rivalry of other men can have far more children than those who fail. Because of the privileged access that awaits the successful men, in fact, many of their defeated rivals will fail to father any children at all. It's hardly surprising that the urge to mate should have come to be so dangerously insistent in almost all men: after all, everyone on the planet comes from a line of males who succeeded at least once. The urgency of that challenge turns every man into a potentially deadly rival to every other.

These pressures on men and on women don't neutralize each other: they're mutually reinforcing. It's a spiral: the selectivity of women encourages the persistence of men, and the more persistent the men, the more selective the women have to be. In other species, we can see this spiral at work in elaborate and sometimes gruesome ways. The water strider, *Rheumatobates rileyi*, has seen an evolutionary "arms race" between evolved resistance in females to male copulation attempts and male armaments to grasp and suppress females who resist. The males have elaborate, hooklike antennae: these are used for holding down females and serve no other known purpose. Or consider the tunnel-web spider, *Agelenopsis aperta:* the male anaesthetizes the female with a powerful toxin and mates with her while she is unconscious. Many male scorpions appear to sting their females during their elaborate mating dance: the scorpion *Parabuthus transvaalicus* produces two types of venom and uses the milder type to immobilize the female for mating. Perhaps the nastiest weapon of all belongs to the bedbug *Cimex lectularius*, which punctures the female's abdomen with a daggerlike projection and injects sperm directly into the body. The costs to the female (in infection risk, blood loss, and organ repair) can be high. But provided she survives to bear the male's offspring, the violence of the male's assault carries an adaptive advantage regardless of the cost to the female's long-term health.

The tango between the selectivity of females and the persistence of males has taken even more elaborate and often more delicate forms in *Homo sapiens* than in these other species. The power of this simple logic is extraordinary. From one basic difference in cellular architecture between the sex cells of males and females—one large and scarce, the other tiny and abundant—and from the asymmetry in investments that males and females consequently make in gestation and parental care, think of all that has followed: the Trojan War, the Roman empire, the sonnets of Shakespeare, perhaps even the whole of our human civilization, founded as it is on the large brains that enabled our reproductively successful ancestors to assert themselves in prehistory. It may be hyperbole to claim that the shape of a single nose can change the world, but it is no more than simple fact that gangs, robber bands, legions, armies and empires, and all of the pomp and show that accompanies them have been built on the lethal competitiveness of men driven by the urge to leave their genetic imprint on the future and by the knowledge that women are the gatekeepers of that future.[10]

Men, like dance flies, have responded to this female scarcity by cornering scarce economic resources of their own. That scarcity has left its imprint on modern society, first through its imprint on the brains and bodies we have inherited from our ancestors, and second through our use of those brains and bodies to navigate our changing natural and social environments. Those environments are spectacularly different from those of prehistory: we live in mass urban societies; we travel great distances and interact with strangers; we use contraception; we talk, write, and fantasize about sex as an art and a game and not just as a means of procreation. We can communicate without seeing or touching one another directly, and we are surrounded by artificial

Male bedbug (*Cimex lectularius*) mounting a female. Note the daggerlike projection at his tail. © PSMicrographs.

or fictitious representations of others (such as photographs and videos) to which we are constantly having to fashion artificial or fictitious emotional responses. Still, even in this utterly artificial environment, every man and woman alive today has emotions and perceptions that are shaped in part by the simple and natural asymmetry between sperm and eggs.

Natural selection can be compared to a tunnel stretching back billions of years to the dawn of life, a tunnel that has shaped everyone whose ancestors managed to pass through it. Sometimes it has been comfortably wide, sometimes painfully narrow, according to the harshness of the environment and the difficulty of the struggle for survival at various times in the past. Since sexual reproduction first evolved many hundreds of millions of years ago, males and females have had to pass through the tunnel together, each fitting into the space left by the other and shaping the space through which the other must pass. You might think that would have left us cramped, squeezed, hardly able to breathe, let alone to move. Yet that tunnel has opened out dramatically in the modern world. Humans live in natural and social conditions that are extraordinarily diverse and mostly very different from those in the African woodland savanna in which we first evolved. We may feel cramped by our passage through the tunnel, but now we can stretch our legs, breathe the fresh air, and start to move around. Just as intriguing as the way our sexuality has been shaped by its long passage through the evolutionary tunnel is the way it has begun to adapt to our new and more spacious social world.

During the passage through that evolutionary tunnel, men who could acquire economic resources were able to coerce or bribe their way into sexual reproduction and left more descendants than those who could not. Those conditions have changed beyond recognition today, but if we are to make economic inequality between men and women a thing of the past, we need to understand the psychological marks that the tunnel has left on us. The fundamentals of our inherited sexual psychology are simple, but the details are subtle and often very surprising.

The Impact of Female Scarcity on Human Psychology

Throughout our evolutionary history, women's sexual psychology has been shaped by the need to be selective in their reactions to men, just as men's has been shaped by the need to be persistent in their approaches to women. It's as simple as that. The reason we may not understand this process is that while some of the negotiations between men and women are conducted at a conscious level, many are conducted through the operation of our emotions and our instincts, which influence our choices without our realizing exactly how they are doing so. These emotions were fashioned by natural selection in the physical and social environments of the late Pleistocene era, which in some ways were very different from those of today. Although those environments selected for a psychology that was remarkably flexible compared to that of other animal species, its flexibility was not limitless. We are navigating the twenty-first century AD with instruments from before the twenty-first millennium BC.

Beyond these simple truths lies a whole landscape of varied and surprising consequences, in which many of the stock generalizations of folk psychology (of the "Men are from Mars" kind) fail to hold up. Being selective in their reactions to men is compatible with a large repertoire of psychological responses to the many different situations in which women find themselves. Consider the traditional idea that men are incorrigibly promiscuous, women fundamentally monogamous. It's not clear how many people ever really believed it, as opposed to wanting to believe it. At any rate, it gains scant support from biology today. In the early days of evolutionary psychology, and even well into the 1980s and 1990s, some researchers drew hasty conclusions from the contrast between female selectivity and male persistence. Men's greater reproductive interest in quantity, it was said, meant that promiscuity was inherent in the male brain; selectivity was similarly assumed to imply an instinct for monogamy in women.

Such conclusions led to great controversy, and evolutionary psychologists were often accused of peddling a reactionary and sexist agenda. This was partly because they seemed to endorse an unflattering Victorian picture of women as passive creatures of limited libido,* and partly because they appeared to endorse the common double standard that condemns infidelity in women while condoning it in men. Those who disliked such conclusions were often driven to reject the very idea that our psychology might have been significantly shaped by natural selection or that natural selection might have operated differently upon women and men. This is invalid reasoning: natural selection has no interest in either flattering or demeaning us or in justifying or condemning our common patterns of behavior. There are good reasons to think that natural selection has shaped such admirable human traits as altruism as well as such deplorable traits as our capacity for vio-

* An extreme version of this view was expressed by William Acton in 1857: he opined that "the majority of women (happily for them) are not much troubled by sexual feelings of any kind" (Acton 2009, 112).

lence. Indeed, one plausible theory even suggests that the two kinds of trait evolved together, each helping the other along.[11] So the way to judge a theory about our evolutionary origins is certainly not according to whether it makes us feel uncomfortable (a trap into which many of Darwin's original critics fell). But even if it were justifiable to reject a scientific argument on the grounds that we dislike its conclusions, the view that natural selection has made men promiscuous and women monogamous is factually incorrect. That in many species males vary greatly in the number of their sexual partners may come as no surprise. But the same is true of females.

Thanks to careful animal observation in the wild, to DNA studies, and to more careful reasoning about the evolutionary logic of various patterns of behavior, we now know that females of many species, including many birds and mammals, are sexually adventurous, have high libido, and are often far from sexually monogamous even when they live in socially monogamous pairs.[12] This doesn't make them undiscriminating: in group-living species, there are usually enough available males to allow a female to have multiple sexual partners while still being highly selective about who those partners are. Nor does it mean that female sexual appetites are just like those of males: on the contrary, while males and females may both display clear enthusiasm for high-quality sex, females are more likely to prefer no sex to mediocre sex, while males of many species are quite happy to accept mediocre sex if the alternative is no sex at all.[13] But there is nothing shy or inhibited about the profile of female sexuality that emerges from current work in animal behavior and in evolutionary psychology. If anything, it is even more complex and Machiavellian than that of male sexuality, since it reveals just how much females have to gain from deception.

The benefits to males from deception are not quite as great, for an interesting reason that has its origin in the way females are obliged to care for their offspring while males are not. A male who has sex with many females within a short period of time can potentially have many viable offspring. Each of the females has an interest in contributing to the care of her own offspring, whatever she may know about the habits of the father. It doesn't pay a male to make too much effort to cover his tracks after the event, from the females at least (the rival males are another story).

Things are different for a female who has sex with multiple males within a short time; this pattern may have been fairly common for human females during prehistory (see chapter 4). She may gain contributions from these different males, some in the form of DNA and some in the form of food or protection. But there will normally be only one child, if any, from closely spaced sexual encounters, and only one of the males can be its father, so that the reproductive interest of multiple males in contributing to nurturing the fetus and raising the child will depend on uncertainty on their part as to which of them is the real father—a confusion the female can only benefit from encouraging.[14] Coyness may not be an inherently female strategy, but it's sometimes in the interests of females to make their males think it is, especially in response to the sometimes violent jealousy of males. If Victorian moralists were under the impression that women's sexual appetites were inherently limited, that shows only how effectively the wool had been pulled over their eyes by countless generations of women.

If monogamy is often not what it seems, the alternatives to monogamy can take strikingly different forms in different settings. The sheer variety of sexual behavior across animal species, as well as across human societies, makes it difficult to generalize about how natural selection has affected relations between the sexes. Who would think, for instance, that human males are undiscriminating in their sexual pursuits while women are selective, when the cosmetics and fashion industries worldwide make sales of several hundred billion dollars every year largely on the promise of making women more attractive to men? But there's a reason for all that variety between species and between societies: in every animal or human setting, males and females develop their strategies, consciously or not, in the context of the environment created for them by the strategies of the other sex. It makes sense to drive on the right if everyone else is doing so but to drive on the left if everyone else is doing *that.* Similarly, the sexual strategies of females in any species are tuned to the strategies of the males of that species and vice versa. Otherwise-similar countries have different rules about which side of the road to drive on and conventions about how the sexes should behave to each other (men and women can exchange playful and flirtatious glances on a Parisian street, but identical expressions might be considered lewd, aggressive, and offensive on a street in Washington, DC).* Otherwise-similar species (like gorillas, chimpanzees, and bonobos) have arrived at very different models of relations between the sexes. But that doesn't make it redundant to try to understand the evolution of sexual behavior in terms of adaptation to an environment: instead it reminds us that we have to include in that term the conditions created for each sex by the behavior of the other.

In the evolution of the human species, males found ways to compensate for the cheapness and abundance of their sperm. One of the most powerful ways to do that was to corner the scarce economic resources in their societies, for which women learned to compete in their turn. This gave the relations between men and women an intriguing twist that was quite different from what we see in most other animals, including our close relatives, the great apes. How did these relations develop, and what do they mean for modern societies, where access to economic resources is much less unequal than it was in the environments where human beings evolved?

The Structure of This Book

This book asks how an understanding of our biological inheritance can cast light on the forces shaping relations between men and women in the twenty-first century. Chapter 2 focuses on that biological inheritance, surveying what we share with other sexually reproducing species in general and with other primates in particular. It explores in particular how signaling, and the opportunities for manipulation that come with it, dominate courtship activ-

* There are many other intriguing cultural differences between France and the United States: to take just one, French has no equivalent for the English *date*, meaning a meeting between potential lovers at which they size each other up. In the French language, it appears that sizing up occurs whether or not there is a prior agreement to do so.

ity. It looks at the different strategies used by males—the impoverished sex, launching their gametes hungry into the world—to manipulate females, whose gametes come with a dowry of food and protection and who have to be very choosy about whom they share that dowry with. And it looks at the strategies used by females to manipulate males in their turn.

Chapter 3 explores the way our emotions interact with our conscious rationality in sexual signaling. The strange thing about our emotions—the fact that we have so little conscious control over them—turns out to be a strength in signaling our trustworthiness to potential sexual partners. It allows us to make more solid commitments than would ever be made possible by the use of rational calculation alone. Much of the elusive, infuriating, and enchanting nature of sexual courtship comes from the way we hide from ourselves the true nature of what we feel and why we feel it. Far from being a flaw in our makeup, it is a testimony to the complexity of the problems natural selection had to solve to enable us to handle sexual reproduction at all.

Chapter 4 looks in detail at our primate inheritance, at how we have expressed it, and how we have changed it. This chapter explores what has made us different from other primates and looks in particular at the ways in which our ancestors used resource scarcity as weapons in the sex war. Humans are unusual among primates in having young that take many years to rear; this meant that our female ancestors needed the resources of group living, including the economic resources contributed by males, to reproduce successfully. This dependence on males had its price, for men became more possessive about their children and the women who bore them. Whether men's possessiveness was felt as a minor nuisance or as an oppression depended on the physical and economic circumstances of the societies concerned: farming communities could bully and confine their women much more than hunter-gatherers, for instance. And one extraordinary long-term consequence of this shift to intensive child rearing was that it gave human beings the opportunity to develop large brains that made them flexible and adaptable, able to transcend the behavior patterns that had evolved for one environment and refashion them to suit another.

Chapters 5 to 7 look at relations between men and women today, when the social and economic conditions in which we live have changed beyond all recognition from those of the hunter-gatherer communities in which our brains evolved. We start by looking at the great experiment of gender mixing that took place in the twentieth century, when formal barriers to women's participation in almost all areas of economic life were removed in most of the industrialized countries, and the gender division of labor that had existed since prehistory collapsed under a tide of talented and energetic women who moved into large areas of economic and social life previously monopolized by men. Why did this tide not reach everywhere? Why have relative incomes for women stagnated at around 80 percent of those of men, and why have some occupations and positions of power remained so persistently masculine when others have become so rapidly and uncomplicatedly mixed? Chapters 5 and 6 look at two possible explanations: differential talent and differential motivation. Neither one can really make sense of the facts. In particular, though there are indeed differences on average between men and women in their talents and aptitudes, it's remarkable how small these differences are and how inadequate they are to explain the large differences that still persist in the representation of women in positions of economic power in modern societies. Women do have different preferences from men, on average, and they do have somewhat different aptitudes (again on average). The puzzle is how high an economic price they still seem to pay for these differences, a price that seems inappropriate to the needs of the modern world.

Chapter 7 puts forward a different explanation. It looks especially at the different ways in which men and women form coalitions and networks, an activity central to the life of all primates that live in groups. We all build around ourselves a web of contacts and affinities that bring us closer to some of those on whom our happiness and prosperity depend while simultaneously distancing us from others. It seems that the lack of congruence between women's and men's networks makes it harder for women and men to interact on equal terms. While men's networks play an important role in giving them access to positions of economic power, there's evidence that women's networks don't have the same effect. Whether that's because of how women deploy their networks or how men respond to them, or to a subtle interplay between the two, is hard to tell. At all events, it's in these more subtle interactions that we can find an explanation for why such apparently small differences in preferences and aptitudes can lead to such persistent exclusion of women from positions of economic power.

Chapter 8 sets out to look at human cooperation more widely. Cooperation and conflict between the sexes may have a particularly urgent tug on our intellect and our emotions, but the whole edifice of modern life is built on similar challenges. However much technical progress may expand the economic possibilities for humankind, realizing those possibilities will require us to solve some large and primitive obstacles to effective cooperation. In particular, modern working life is not just about learning how to function efficiently in concert with technology; it also involves sorting ourselves into groups that work well together. The chapter discusses how technology can help us do this sorting more effectively, but it also points to some fundamental limits to what technology can do.

The greatest predicament of our sexual lives is the fear—sometimes well founded, and often corrosive even when it is not—that those we want as partners may not want us in return. And this predicament—exclusion from cooperation with those whose cooperation we most value—recurs in many ways throughout our social and economic lives. It's certainly not confined to women. In fact, in parallel with the puzzle about why women are so systematically excluded from positions of economic power, there's a puzzle about why men are so heavily overrepresented not only in the top echelons of society but also among the undereducated, the unemployed, the homeless, and the incarcerated. This so-called crisis of men is sometimes thought to undermine the complaints of women about exclusion. But this is a mistake. Both the prevalence of men at the bottom of society and the exclusion of women at the top are cause for concern. And a broader, more sobering conclusion is that whatever the promises of information technology for enabling collaboration in our social and professional lives, nothing is going to solve the predicament of those with whom no one wants to collaborate. The most sophisticated matching methods can identify for us the most likely willing partners for social, sexual, and economic partnerships of all kinds. No technology in the world can ensure that those partners really are willing to cooperate with us.

Chapter 9 draws conclusions from these findings that can help guide our ethical and political choices. There are good reasons to think that economic change is undermining the conditions that made possible the massive subjection of women in traditional agricultural societies. So if an understanding of biology can sometimes make us feel we are the victims of our inheritance, an understanding of economics shows that we can adapt that inheritance to shape our future. Our ability to do so is due to the sensitive interplay between the bodies and minds that we have inherited from our ancestors and the wide range of natural and social environments in which our species has learned to live. The human species has learned to use its large brain to refine its survival strategies in an extraordinarily flexible and creative way. This capacity doesn't mean that anything is possible, because large brains are expensive to build and maintain, and natural selection has taken risky shortcuts in designing them—shortcuts that limit how we can live. All our emotions and desires are shortcuts that allow humans to economize on expensive brain tissue and steer us in directions that have proved advantageous for us in the past. Our taste for sugar, for instance, was a reliable guide to adaptive eating in the Pleistocene era, but in the very different conditions of modern life, it threatens us with obesity and diabetes.

But if biology teaches us which of those shortcuts men and women have inherited from our ancestors, economics invites us to admire how intelligently and flexibly those shortcuts can be put to use in modern environments. The economic circumstances in which we live today are much more favorable to a resolution of many of our conflicts than was true for much of the recent past, though we may need to engage in some ingenious manipulation of our emotions and instincts in order to achieve such an outcome. Learning how our emotions and instincts are constructed helps us to navigate more flexibly in the world we have created.

But the message from biology is not entirely reassuring. Sex has evolved through the clash of conflicting interests, and it remains a turbulent area of our lives, with few easy rules to follow. Each of us is descended from innumerable generations of men who lied, cheated, charmed, bullied, or killed their way to sexual intercourse, and from innumerable generations of women who charmed, seduced, lied, or manipulated their way to extracting economic privileges in return for access to their bodies. All of those men and women have planted their seed in us: it's hardly surprising that its flower in us, their descendants, should have some trouble growing straight. Everything they did, all their ambitions and their dreams, would have left no trace on us had they not succeeded in reproducing, and we will not understand those ambitions and dreams unless we understand the sexual contest that shaped them.

If the diagnosis is not as reassuring as we would like, realistic prescriptions may also have to be more modest than we might hope. Placing our faith entirely in the law to remove discrimination and inequality will make reformers grow melancholy while bringing smiles to the faces of their attorneys. The law is part of the solution, but it needs to take account of the complex causes of current injustice. Powerful desires, emotions, and attitudes can often simmer inside us, confounding the efforts of respectable society to tame them. Nor should we expect too much from a policy of recommending honesty and openness in all situations: sex is too rooted in deception, exaggeration, and manipulation for us ever to be entirely straightforward about it. If we clamor for openness and pretend that it is attainable, we may be ill prepared for navigating the shadows that are inevitably cast by the sexual life. A more modest hope is to be lucid about, and unafraid of, the confusions and deceptions that we are certain to encounter.

If men and women were negotiating entirely consciously, we could at least be lucid about our disappointments. But when our emotions are guiding the transactions for us, the disappointments hit us with the full force of the unexpected. His twenty-first-century rational brain can see that she likes him and wonders why she resists his sexual overtures (so much pleasure, so little cost), not understanding that her emotions may be sizing up the proposition as it would have appeared to her in the Pleistocene: a massive commitment, fraught with risk. Her twenty-first-century rational brain wonders why sex is such a big deal to him, not understanding that to his Pleistocene emotions it's not just a big deal—it's the only deal. His rational brain wonders why, if she's not interested in sex, she should get so upset if he starts to show a sexual interest in someone else. Her rational brain wonders why, if he's so relaxed about his wandering appetites, he should become suddenly so anxious if her appetites start to wander too. Their hormones are filling in the scenery for both of them with the sounds, the scents, the fears and dreams, the whole emotional meteorology of the Upper Paleolithic, when sexuality was not just a matter of lifestyle but of life and death.

Lucidity is hard to come by, too, in social and public life. It is vain to hope that whole areas of public life—politics, say, or the workplace—could one day simply operate as though we were asexual workers and citizens. Biology warns us clearly that the average behavior of women will not grow to be just like the average behavior of men (fortunately so, given men's greater disposition for violence). Men, on average, will continue to want different things from what women, on average, want. Men and women will continue to use different strategies to pursue their ambitions. All individuals, men and women, will also want contradictory things: to be successful and to be protected, to choose our partners and to be chosen by them, to be passionate and to be reasonable, to be forceful and to be tender, to make shrewd choices and to be seduced. With such contradictory impulses, all of us will sometimes make choices we regret. Sex is about danger as well as about tenderness: the two are inseparable, and they are what has made us such a tender and dangerous species.

TWO

Sex and Salesmanship

> She must be soaping herself at this moment, he was thinking in the bath. Though he yearned to see her soon, he could not help feeling how ridiculous were these two human beings, at the same moment three kilometres apart, rubbing and scraping as though at so many dirty dishes, each in order to please the other, like actors preparing for a scene.
>
> —Albert Cohen, *Belle du Seigneur*, 1968

Signals on the Street

THIS IS AN ADVERTISEMENT FOR DNA. A consignment of the world's highest-grade deoxyribonucleic acid is headed this way, in twenty-three pairs of subconsignments, two of them making up the XX combination that I admire so much. Better still, some of it has undergone meiosis and is ready to play its part in reproduction. Here, advancing along this busy street, the consignment appears in its full gift wrapping—dazzlingly symmetric, as if to show off the perfect conditions in which it was produced, or perhaps the fact that the blueprint is robust enough to survive even an unhealthy gestation without conceding the tiniest flaw in the architecture. It comes toward me, swaying slightly to emphasize its perfect balance and to accentuate those hips that will launch another generation of genes. The designer fabric in which the whole package comes expensively wrapped pretends to be concealing and protecting the contents while actually flaunting them outrageously to the world.

She smiles, this audacious triumph of genetic marketing, and a computer buried in the gray porridge in my skull starts to run down the checklist before I am even aware of it. The muscles crinkle at the corner of her eyes, showing that she's feeling relaxed and friendly, predisposed to cooperate.[1] Her teeth are sharp, strong, and gleaming with health. The skin is flawless—

no parasites; no lurking recessive mutations that I can spot: an immune system to die for, or rather not to. The eyes sparkle—not the faintest cloud on the cornea. The hair shows the luster of perfect nutrition. Those graceful legs look as though they can dance to the most complex rhythms, balance on the most treacherous terrain, and probably outrun most predators, including me.[2] Her tint is a little darker than mine, her cheekbones attractively flared, her fugitive aroma enticingly spiced—or am I imagining this? She embodies enough exoticism to leaven my insipid genome, but not so much that her kin and mine couldn't rub along.[3] It's a cocktail whose ingredients stimulate more neural receptors than I ever knew I had.

I want some of that DNA badly: we could make some very fit descendants together. At least, that's what the scientific part of my brain tells me, but she's pressing some buttons in me that the scientist doesn't seem to control. All effective advertisements draw a veil over their true purpose, of course. She may be advertising her genes, but to me, suddenly breathless under the impact, it doesn't feel as though it's her genes I'm after at all.

I could invest a lot in you if you'd only let me, my eyes implore her. To no avail, of course. Her advertisement was not really aimed at me—except in the faintly condescending way common to advertisements for diamonds and sports cars that I shall never buy—but at an altogether superior target group of men. Beside her, fingers entwined in hers, strides another advertisement, not aimed at men at all, except to warn us to stand aside. He looks superbly fit and effortlessly superior to the likes of me. These two world-class marketers have found their respective target audiences, and they match—for now. It's a poignant moment, but the computer in my brain won't allow me to linger in regret. Already, as she looks past me, I'm looking past her. The streets of this magnificent city are humming with some of the world's finest DNA, and there will be some more along in a moment.

Sex Needs Advertising

Sex is inseparable from advertising. Every member of a sexually reproducing species is built to advertise and to cast a critical eye on the advertising of others. In many species, though, the advertising budget is a modest portion of the organism's overall investment in survival and reproduction. Sometimes that's because finding sexual partners is easy, particularly for females: there may be lots of males around, and they may be so easy to attract that it's not worth investing in much more than a basic signal that you're alive and in reasonable health. Sometimes, and much more often for males than for females, fighting for sexual favors makes more sense than advertising for them—though advertising to impress rival males can sometimes persuade them to back off without a fight. But in many species, at least one sex may devote enormous resources to advertising for the favors of the other: the spectacular tail of the male peacock, the elaborate nests of male bowerbirds, and the songs of male nightingales are just some of the better-known examples. In most advertising species, one sex displays, and the other responds: peahens are famously drab, female nightingales notoriously unmusical. One of the most unusual things about human beings is that advertising is a preoccupation—one might almost say an obsession—of both males and females. Why?

A clue to the answer comes from the courtship behavior of the dance fly that we met in chapter 1.[4] Male dance flies compete for females, as do males of all species, but they use ingenious methods that provoke an unusual response. Female dance flies gather in swarms. The males compete by capturing smaller flies or arthropods as prey. They enter the swarms, offering these gifts of food to receptive females, who feed while mating.

This simple shift in the male strategy of seduction has one remarkable consequence. Since the best males are not ostensibly offering their sperm, which is plentiful, but their prey, whose supply is limited, the females compete with one another for this scarce food. A male who bestows his favors on one female has less to offer another, and it therefore pays for females to make a serious effort to be at the head of the line. So the females resort to a strategy that is more usually found in male animals: they advertise their fitness. But their advertisements are deceptive. They have evolved hollow sacs of air on their abdomens that they inflate with air to mimic the look of a fertile female full of eggs. This strategy works because males are attracted to signs of female fertility. In a similar way, silicone breast enhancement in human females really does work to attract (some) men. In the past, because large breasts really were associated with fertility, men's hormonal systems were trained by natural selection to respond to this cue. Humans use expensive silicone while dance flies use cheap air, but the principle is the same.

Females are not the only ones to engage in deception. The male dance flies usually wrap their food offerings in silk, and (as human beings know well) a silken covering can hide anything, or nothing. Sometimes the packages are empty. Obviously, if they were always empty, natural selection would not have favored females that were drawn to them. Females who are seduced by food and silk have been favored by natural selection because, overall, food and silk really do offer opportunities for enhanced fitness.

Human sexual signaling is more elaborate than the dance fly's. Men can turn almost anything into a sexual signal, from the cars they drive to their conversational style, from generosity with jewelry to virtuosity in a jazz solo. The qualities they may be advertising are similarly varied, including intelligence, healthy genes, kindness, fidelity, and a talent for dealing with children. Women, in turn, use a vast array of signals in clothes, cosmetics, conversation, dancing, and songs to advertise an equally wide range of qualities.

To complicate matters further, not all the advertising humans do is aimed at potential or actual mates: we also advertise to colleagues, friends, enemies, clients, and business partners or competitors.[5] The methods we choose are astonishing in their subtlety and duplicity: many hygiene rituals, for instance, signal concern about hygiene while doing nothing to reduce infection risk and sometimes even increasing it.[6] Some of our advertising to these other parties is observed by potential mates, who have reason to care about how well we succeed in attracting friends and defeating enemies.[7] Often, faced with multiple audiences, we find ourselves toning down the advertising directed toward some people in order to avoid sending counterproductive signals to others. Erving Goffman calls this practice "covering" when minority groups try to make their identity seem inconspicuous without actually trying to hide it, and Kenji Yoshino has more recently called attention to the relentless pressure for covering faced by homosexuals, ethnic minorities, and many other minority groups, including women working in traditionally masculine settings.[8] The signals we send are not directed lasers: they fan out through our social worlds, provoking recognition, suspicion, excitement, curiosity, and alarm. In the cacophony of claims for attention, how can anyone interpret what these signals really mean? What makes the signals reliable indicators of anything at all?

Advertising Needs Credibility

The dance fly shows us that even when advertising is deceptive, it is rarely outright false. Sacs full of air are deceptive, but they continue to attract males because, on average, females with enlarged abdomens really are more fertile. Silken packages may sometimes be empty, but they continue to attract females because, often enough, they contain scarce food. The advertising we see on street hoardings is never completely honest—it always exaggerates a product's quality, style, or reliability. But rarely does it tell blatant lies, for to remain persuasive it must also be plausible: if the products advertised on billboards or television were systematically no better than their rivals, we would learn to avoid them in favor of products that were cheaper because their advertising budgets were lower. A better way to express it would be that advertising is disciplined manipulation, held to standards of at least partial honesty by the requirements of credibility, but manipulation all the same. Advertising seeks to influence us in ways that are in the advertiser's interests. But because we can respond intelligently, the manipulation has to respect certain rules to be effective, and those who are manipulated can manipulate the advertisers in turn. Advertisers are not required to speak the unvarnished truth, but their pitches have to make sense to us, or else the effort is wasted.

The history of commercial advertising gives us plenty of examples. Among the first advertisements to reach a mass consumer audience were those for patent medicines in the United States in the nineteenth century.[9] The shockingly poor quality of many patent medicines quickly educated the reading public against believing everything they read in the newspapers, or at least in the advertising sections. The copywriter John E. Powers, known as "the father of creative advertising," realized that advertisers needed to regain a reputation for honesty if their pitches were to have any value. He once wrote an advertisement for a Pittsburgh clothing company that read "We are bankrupt. This announcement will bring our creditors down on our necks. But if you come and buy tomorrow we shall have the money to meet them. If not we will go to the wall."[10] The result was an immediate surge in sales that saved the struggling company, presumably boosted not by altruistic customers but by those who sensed the opportunity to snap up a bargain. Similar motives probably underlie the popularity of songs in which the singer laments being dumped by a lover (when did you last hear a singer sing about dumping someone else?). The song aims not to draw attention to the singer's failure in the love market but rather to make a show of honesty that advertises her sudden—and preferably temporary—availability in order to provoke a surge of interest from new suitors. An even more subtle piece of manipulation comes in the form of songs advertising a woman's faithful adoration of an inadequate or abusive man: I'm not really unshakably faithful to *him*, the song seems to hint, but if you were a bit better than him, I could be unshakably faithful to *you*.

The tango between advertisers' manipulation and consumers' skepticism has continued down to our own day. Television commercials for washing powders from the 1950s now seem to us charming in their naiveté, for their presumption that a viewer will simply believe an actress pretending to be a housewife who says that her favorite powder really washes whiter than some rival brand. Later advertisements progressively revised their pitch. They replaced housewives with white-coated scientists until viewers realized those were just actors too; then they developed scenarios so sophisticated and expensive that the baffled viewer had to conclude that the powder must be good or else no one would have gone to so much trouble or spent so much money to promote it. An advertisement for mid-range cars a few years ago featured the supermodel Claudia Schiffer, whom most viewers would hardly associate with having much mechanical expertise and certainly not

Supermodel Claudia Schiffer signaling to the public the merits of the Citroen Xsara car, 1998. © Yves Forestier / Sygma / Corbis.

with driving anything less expensive than a Porsche. However, viewers do know that she is spectacularly well paid. They might therefore reasonably conclude that the manufacturers must think their product good enough to be worth backing with a very large budget. The advertisement may be wasteful, but its wastefulness is precisely the point: a more prudent advertising campaign would be a less enthusiastic signal of what the manufacturers believe about their product's worth.[11] At each stage in the escalation of tactics, the advertisers know that viewers cannot be fooled forever: a successful campaign makes the viewers in some sense knowing accomplices of their own manipulation.

Not all advertising has to worry about credibility. A poster giving the telephone number of a public helpline, a street map displayed by the city's tourism department, or a plumber advertising twenty-four-hour service can take a low-key approach: nobody doubts the reliability of the information because the advertisers have nothing to gain from lying. In a similar vein, some of the displays that people routinely use are proof against skepticism because lying seems so pointless. Why sport a bumper sticker declaring yourself to be a Republican if you're really a Democrat, or wear a Manchester United scarf if you really support Chelsea? When the point of signaling is to communicate with others who share your tastes, deception yields no benefit.

But such clear-cut cases are rare. Suppose I sport a discreet badge saying I belong to a yacht club. Am I merely signaling my tastes in a neutral fashion so that I can meet up more easily with others who share them, like those who support the same football club as I do? Or is it designed also to impress people, yachtsmen and landlubbers alike, who know how expensive a hobby sailing is and will therefore presume me to be prosperous and successful? In that case, they also have to believe that I didn't just buy the badge for a few cents in a corner store. After all, nobody believes you went to a prestigious university just because you wear its T-shirt, given that such shirts can be bought for a few dollars on a street corner in San Francisco or Shanghai. The badge has to be hard to acquire—either available only to members of the yacht club or so expensive to fake that the only person who would want to do so would be someone rich enough to belong to the yacht club anyway. But, paradoxically, that means that the people who have to work hardest and spend the most to project themselves as yacht-club members are the true members themselves: if the signal is a credible one, others won't bother. As Cary Grant famously put it: "Everyone wishes to be Cary Grant. Even I wish to be Cary Grant." His talent and passion for self-impersonation have become the stuff of Hollywood folklore.[12]

The anthropologist Geoffrey Miller has written amusingly and illuminatingly about the kinds of signals people send about their personality types.[13] People who are conscientious, for instance, often like to hang out with other conscientious types (and those who aren't conscientious typically don't). Does that mean that sending signals of your conscientiousness is just a matter of coordinating with others who share your tastes, like sporting a political bumper sticker? Not quite. The problem is that being conscientious is hard work, and most of us would benefit from being thought to be more conscientious than we really are. That is all the more important if we are trying to establish a relationship in which our conscientiousness really counts.

So a truly credible signal of conscientiousness has to be costly, otherwise everyone could pretend to be conscientious whenever it suited them. Have you ever wondered why some people lavish attention on rare breeds of dog or cat that have delicate health and require an enormous amount of money and attention? Miller's book will tell you that the cost is not an incidental disadvantage of owning these pets: it's the whole point. It is a credible signal to the outside world that you are an unusually conscientious type of person. The easiest way for a single man and a single woman to meet and strike up a conversation in the street is by walking their dogs. The rarer the breed, and the more demanding its upkeep, the clearer the signal that the owner is a conscientious person. Of course, an occupational hazard of meeting a conscientious person is that you may end up having to help look after the dog, just as an occupational hazard of being Cary Grant is how hard you have to keep working at being Cary Grant.

The Need for Credibility Encourages Extravagance

The idea that the real message of an advertisement might sometimes be its sheer extravagance rather than its specific content (as in the car advertisement featuring Claudia Schiffer) goes back a long way. Thorstein Veblen, in his *Theory of the Leisure Class*, developed the notion of "conspicuous consumption" to describe the motivation of some rich people to show off the sheer amount of their spending rather than the specific things they buy.[14] The French historian Paul Veyne developed the idea to explain why Roman emperors notoriously used "bread and circuses" to pacify the citizens, even though the citizens were being bribed with paltry sums in comparison to the costs of their own subjection and were usually being bribed with their own money to boot. The explanation, Veyne suggested, was that the citizens were not being bribed: they were being threatened. The signal sent to a grudging and unruly population by lavish expenditure on bread and circuses was not "The emperor loves you," but "The emperor is so rich that he can afford to throw money away and still have more than enough to hire as many mercenaries as he needs to put down the most carefully organized rebellion."[15]

Examples abound in history and legend. When the French fortress town of Carcassonne, then a Saracen stronghold, was under siege by the emperor Charlemagne in the ninth century, it is said that its queen, Dame Carcas, had the idea of gorging a pig on the town's meager remaining supplies of grain and throwing the body over the battlements. As she intended, Charlemagne concluded that a city that could afford to feed its pigs on grain must have enough stocks to see out a very long campaign, and he abandoned his siege, later accepting a peace treaty on Dame Carcas's terms.*

Many of the most wasteful aspects of our modern consumerist lifestyle are hard to understand if you think of their wastefulness as simple oversight, as though it had not occurred to us that we could live less extravagantly than we do. It's common, for instance, to hear calls for investment in green technologies to reduce the fuel consumption of cars. Yet almost every car owner could reduce fuel consumption, with no new technology and without driving less, by the simple expedient of buying a less powerful car. Buying more powerful and expensive cars than we need is a signal to others. The challenge for innovation is not so much to develop technologies for producing green cars more cheaply as it is to come up with ideas for producing cars that are expensive enough to serve as status signals for the successful without causing environmental damage. As things stand now, even Hollywood stars who preach environmental restraint tend to buy ecoconscious hybrid cars to add to their SUVs rather than to replace them, as the journalist Jonathan Foreman reported in November 2009.[16] In the words of a Hollywood agent, the Toyota Prius is "the only car you can drive which costs under $35,000 which doesn't make everyone think your career has gone down the toilet." The Prius has a particular cachet because of its distinctive looks: hybrid versions of other cars tend not to look very different from the traditional gasoline models. The economists Steve and Alison Sexton have esti-

* Dame Carcas's action counts, of course, as an instance of successful mimicry of a costly signal, mimicry that works only because the signal is sufficiently costly to be usually credible. Bluffing all-in at poker is another example. For more on this anecdote, see "Dame Carcas," Carcassonne Town Hall and Tourist Office, www.carcassonne.org/carcassonne_en.nsf/vuetitre/DocPatrimoineDameCarcas, accessed May 18, 2011.

mated, using demand data, that Prius customers are willing to pay between several hundred and several thousand dollars more in order to be seen driving a car that can be instantly recognized as environmentally responsible.[17]

Foreman's article also notes the many private aircraft owned by the stars who have been most outspoken about environmental issues, from John Travolta (who owns five jets, including a Boeing 707) to Harrison Ford (with one jet, four propeller planes, and a helicopter).[18] So long as a private jet is a signal of a successful Hollywood career, exhortations to tread inconspicuously on the Earth will never work. Saving the planet has to seem not only cool but really expensive: if it's a gesture anyone can afford, it will never catch on in Hollywood.

One of the great mysteries of executive life in modern corporations is how excited many successful executives become about first-class plane travel. There are some remarkably intelligent and successful people whose conversation is filled with stories about little else (I know a few). If your idea of international travel is based on the color supplements to the weekend papers, you might imagine that traveling executives live pampered lives, stretched out on sun loungers in quiet resorts in outstandingly beautiful places. The remarkable thing is that almost none of them do; the haunts of these upwardly mobile achievers are quite different from those of the horizontal rich. At their foreign destinations they have no time to see anything but hotels and offices that are just like the hotels and offices in their own countries. The most sordid secret of all is that a long-haul flight in first class is noisier and more uncomfortable than a night spent in a moderately well-run two-star hotel; and unsurprisingly, given the conditions in which it is prepared and served, the food is no better than in a mediocre restaurant. Yet frequent first-class travelers recount their past travels with glee and look forward to future ones. The reason should by now be clear. No two-star hotel would charge you one dollar every five seconds for your night's sleep, and no earthbound restaurant would expect you to pay hundreds of dollars for a reheated meal you eat on a picnic tray attached to your seat.[19] Well-managed companies buy first-class tickets for favored employees because it's better that the chief executive negotiate a contract after the kind of night's sleep you might enjoy in a two-star hotel rather than one you might have in a bus station. But the employees themselves enjoy the experience for reasons that have little to do with comfort. As the attentions of the obsequious cabin staff never cease to remind them, it's all about projecting how successful you are, and paying thousands of dollars for a mediocre night's sleep signals success in a way that few other experiences can.

If private jets and first-class travel are not just private indulgences but also, and more insidiously, signals of professional success, what are the signals that we use to communicate sexual availability and success?

Advertising and Seduction

That commercial advertising uses the techniques of seduction is a cliché, as is the idea that sex and advertising are the world's two oldest professions (the earliest known commercial logo is a carving on the walls of Pompeii advertising the way to a brothel). But the explanation for this fact is an important insight of evolutionary biology: seduction itself is a form of advertising. It is advertising about biological fitness, or, to be more exact, about how much the seducer can contribute to the evolutionary fitness of the seduced. Male song sparrows, for instance, have a repertoire of between five and fifteen song types, and female sparrows become visibly more excited by males with a larger repertoire.[20] Patient human researchers have discovered what the female sparrow's brain in some sense already "knows," which is that males with larger repertoires contribute more to the females' reproductive success: they have more offspring, and those offspring in turn have more offspring themselves.[21]

Why should the song repertoire be associated with fitness? It's not as though singing more songs is particularly demanding of energy or stamina: either birds can do it or they can't. Instead, it seems that a larger repertoire is a signal of a more developed brain, since fledglings that face shortages of food show damage to a part of the brain known as the higher vocal cortex and a reduced ability to sing later in life.[22] Nobody imagines, of course, that the female sparrow literally knows these facts about the anatomy of her partner. She just finds him sexy, because her female ancestors were built in such a way that they also found varied singing sexy, because this helped them to leave more descendants, and because she is one of those descendants. Anyone tempted to mock the susceptibility of song sparrows should reflect on the well-known sex appeal of rock stars.[23]

It seems likely, therefore, that a male's song repertoire is signaling to a female sparrow the quality of the genes that she can hope to unite with her own if she allows the male to fertilize her eggs. In humans, seduction is more complicated. Sexual partners have much more to contribute to the partnership than just their genes. Their sexual signaling is correspondingly more complex and more difficult to decode explicitly (though we have all been decoding it unconsciously for at least hundreds of thousands and perhaps millions of years). As potential parents of offspring that require the most elaborate and prolonged postnatal care in the entire animal kingdom—a project that is more likely to succeed if both parties are committed to it over several years at least—human sexual partners are faced by natural selection with a difficult challenge. They're like salespeople who have to convince you not just how good their product is but also how conscientiously the after-sales service and repair staff will treat you if you have trouble with the product later. Human sexual signalers must not only display what's in their genetic shop today but also make a statement of intent about the future, about their plans to contribute to the care of offspring. It's no help that they have somewhat divergent interests in making that contribution, as we've already seen. You might wonder how sexual signaling could ever be capable of signaling any kind of credible promises at all. This is where the emotions come in.

THREE

Seduction and the Emotions

As a little white snake
with lovely stripes on its young body
troubles the jungle elephant
this slip of a girl
her teeth like sprouts of new rice
her wrists stacked with bangles
troubles me

—"What He Said," from *Poems of Love and War*, translated by A. K. Ramanujan (in Ramanujan 1985)

Seduction Targets the Emotions

SEDUCTION IS A PROCESS THAT CAN BYPASS the rational brain, appealing to psychological mechanisms other than those involved in conscious thought. This happens for a good reason. Rational brains are expensive for animals to grow and maintain, so, for many kinds of problem, it is more efficient to rely on solutions from unconscious, emotional brain processes. A song sparrow with a brain large enough to understand the theory of evolution and calculate the fitness-maximizing choice of mate for any situation would be unable to support its vast head on its tiny neck. Instead, song sparrows have developed a small, relatively space-efficient brain that employs shortcuts—simple rules of behavior that enable the sparrow to negotiate the world reasonably well.

Human beings have stronger necks, but the principle is the same. The emotions can be thought of as natural selection's way of embodying some of the necessary rules of behavior. This understanding of some aspects of behavior has come to be known as the "somatic markers hypothesis." Our ability to link our cognitive and emotional responses depends on functions that are now known to be concentrated in specific areas of the brain (notably the ventromedial sector of the prefrontal cortex).[1] There is also a large body of scientific work exploring the role of the endocrine system in mediating social relationships.[2]

Hormones are not the only means by which our brains encode shortcuts: we have well-known, purely cognitive biases, too (like our tendency to see shapes and faces in the clouds), which have evolved under similar constraints. And even our ordinary cognitive decision-making abilities depend on integration with our emotions, as Antonio Damasio (the original proponent of the somatic markers hypothesis) argues in his book *Descartes' Error*.[3] But emotions often program us to make simple but reasonably reliable responses to some frequent and important predicaments with a minimum of scope for reflection. When something frightening happens to me, my body experiences a rush of adrenalin, I feel fear, and I run away (which of these events is cause and which is effect may not always be easy to tell, as it all happens so fast). That might not be the absolutely best response in all situations (I've missed some good movies as a result), but it worked well enough for my ancestors that I am here typing this chapter today.

However, the great virtues of having the emotions direct behavioral responses to our environment—their simplicity and predictability—are also what make them less reliable when that environment changes, as well as susceptible to manipulation by others. A rule of thumb that worked well in the hunter-gatherer environment of our ancestors may be quite unsuitable for our modern environment: finding warriors sexy was once a way for a woman to increase her chances of survival and reproductive success, but in many cities it is now more likely to give her a drug dealer for a boyfriend. More important, a simple rule of behavior that makes us predictable invites others to exploit our predictability. Chief among the manipulators are advertisers: when they realize what kind of advertising we respond to, they make more of it.

Predictability may be a liability, then; but it also facilitates certain transactions. Those who advertise to us are usually not just trying to exploit us but are typically trying to persuade us to do something that is in our interests as well as in theirs (admittedly, on terms favorable to them). Such exchanges can be mutually beneficial, but each party will invest time, energy, and resources in an exchange only if the other party can be relied on. Predictability can sometimes signal reliability—an argument first developed by the economist Robert Frank and now the subject of a growing scientific literature.[4] The display of emotion can therefore persuade others of our trustworthiness in a manner that surpasses the abilities of our consciously calculating brain. Suppose, for instance, that you think I am making friendly overtures to you because I have calculated that it is in my interests to persuade you to trust me. You may respond positively if you think we have interests in common. But you are right to remain wary: any change in our circumstances may upset that calculation, and then all bets are off. If instead you think I am making friendly overtures to you because I really like you, wish you well, and want to spend time with you, you may be much more confident about trusting me. The emotions can be frustratingly difficult for our rational capacity to master, granted; but sometimes for that very reason they have a persistence that is quite foreign to the foxy, calculating brain.

The fact that my emotions are removed from my conscious control is, paradoxically, a strength when it comes to persuading others to trust me. If that's true of ordinary friendships, it's even more true of human sexual relationships, which offer vast gains from cooperation. Those gains will depend on investments made by both parties, and it may not be worth embarking on the relationship at all without an assurance of some commitment. A visibly emotional attachment to you may be a much more credible assurance that your partner will stay around after you've had sex than any amount of reasoned argument on his part. Before the invention of contraception, the risk of pregnancy meant women bore a significant cost in embarking on sexual relationships without some assurance of commitment from men; men, in turn, might balk at such commitments without an assurance of paternity from women. Emotions as signals of commitment could therefore have powerful adaptive value.

Managing emotions so that they are simple enough to be trusted by those who want to cooperate with us but complex enough not to be easily manipulated by those who want to cheat us is a difficult balance for natural selection to get right. And the balancing point would have changed as the arms race between cheaters and honest signalers became more sophisticated over time. The result is that many of our emotional responses are a baffling mixture of the solid and the inscrutable. The solidity explains why we can be faithful and loyal but also why we so often engage in insidiously repetitive behavior with those to whom we are sexually attracted—why we can be so slow to learn that it's time to move on. The inscrutability is the fruit of our continual effort to project ourselves as better (stronger, cleverer) than we are, and it explains both the charm and the frustration of so many exchanges of sexual signals, their playful quality, the fact that nothing can be taken for granted. Why does she give him the come-on only to draw back? Why does he seem to signal his steadiness and loyalty and then behave in completely unreliable ways? Both seem to want to signal their attraction but are afraid of being taken for the wrong sort of ride, afraid of seeming to sell themselves too cheap.

The inscrutability of so many of our sexual signals creates an ideal environment for cultural models to influence our behavior, guiding us toward some interpretations of what others may be signaling to us and steering us away from others. Stendhal describes in *The Red and the Black* the confusions of his young hero, Julien Sorel, and the object of his affections, the innocent Madame de Rênal, the wife of the mayor of a small provincial town in western France whose children Sorel is tutoring:

> In Paris, [their] situation would rapidly have been simplified; but in Paris, love is the child of novels. The young tutor and his shy mistress would have found in three or four novels and even in the poems learned in school the clarification of their position. The novels would have set out a role to play, shown them a model to copy, and Julien's vanity would have forced him sooner or later, whether he felt pleasure in it or not, to follow the model. . . . In a small town in the Aveyron or the Pyrenees, the smallest incident would have proved decisive because of the fiery climate . . . but under these grey northern skies, an impecu-

> nious young man, who is ambitious only because his heart needs some of the joys that money can bring, spends every day in the company of a thirty-year-old woman who is truly modest and busy with her children, and never uses novels as an example to follow. Everything happens slowly in the provinces, it's all more natural.[5]

Novels were considered dangerous reading for young ladies in many nineteenth-century households precisely because even fictional stories could offer previously innocent women a set of methods for decoding sexual signals that might lead them astray. Nowadays, in small provincial towns, even people who read few novels are likely to take their examples from cinema. (Diego Gambetta reports that Italian mafiosi have been highly influenced in their signaling behavior by the *Godfather* movies, so that the mutual imitation of life and art has become in this example entirely circular.)[6] The behavior of courting couples in American romantic comedies and television sitcoms has influenced styles of behavior all over the world.

It's important to understand that the confusion and opacity in the way we signal to each other, and which these cultural models help us try to decode, do not represent some kind of inexplicable dysfunction in the process. On the contrary, the process of signaling has evolved to be confusing precisely because there are many incentives to project misleading signals, and opacity and complexity are the price to pay for signals that can credibly mean anything at all. Transparency isn't a realistic ideal: even the fact that we have nothing to hide can be something we have good reason to hide.*

To see why inscrutability may be a feature rather than a bug in the human emotions, consider what Sherlock Holmes might have called *The Curious Incident of the Female Orgasm*, recalling the famous dog whose significance for the case was that it did not bark. One possible evolutionary explanation for the nature of orgasm in the human female (a phenomenon rare, though not unknown, in other animals) makes its legendary unpredictabil-

* Carrying a camera can be dangerous in authoritarian countries, even in the most unexpected places: the fact that there are no military secrets worth photographing at a certain site may itself be a military secret.

ity part of its point.[7] According to this explanation, the female orgasm developed out of the basic physiology that also enables the male orgasm. But rather than being subject to selective pressures to make it occur more predictably, it remained elusive and unprogrammable as a way of screening males for reliability.[8] This must obviously have happened during a period in our evolution when it was common for women to mate with multiple men (and in which women could exercise choice over their longer-term partner or partners based on the quality of earlier encounters).[9]

Female orgasm, according to this view, came to be associated with hormonal changes (notably the release of the hormones vasopressin and oxytocin) that increase trusting behavior and are associated with emotional commitment.[10] But like all screening mechanisms, it had to be discriminating. If the woman's orgasm could be induced only by the most attentive, considerate, and trustworthy males—the kind for whom it was worth transforming a one-night stand into a longer-term relationship—it could help her to ensure that she didn't make her own emotional commitment too easily and therefore be too easily manipulated. A woman who climaxed with more or less any male she had sex with would be either too vulnerable to manipulation (if the orgasm were accompanied by the appropriate emotional changes) or unable to use her emotions to make commitments (if it were not). From natural selection's point of view, a woman who climaxed too easily would be engaging in the sexual equivalent of grade inflation. Men don't have this problem—but then the diplomas they issue have never been worth much anyway.

It doesn't follow, of course, that the female orgasm had to perform this screening function: other mechanisms could have been found, both for screening and for encouraging commitment, and it may have been an accident that natural selection hit on this particular one. It may have had additional evolutionary advantages.[11] But a physiological mechanism for screening and encouraging commitment would have conveyed a useful adaptive advantage. Unfortunately, though, natural selection has no foresight: a woman's sexual sensitivity may be calibrated to the kinds of men her female ancestors knew, but this notoriously doesn't mean it will be calibrated to the partner(s) she happens to be with at the moment.

A woman's capacity for orgasm may be a form of screening males for reliability, but it is (fortunately) far from being the only skill at her disposal. She exercises other talents long before orgasm becomes even a theoretical possibility. In fact, the more closely we look, the more evidence we find in our daily lives of this interplay of ardent signaling and skeptical screening, with men and women playing both roles. Take smiling, something most of us—even economists—do hundreds of times a day. It's not used only in sexual flirtation, of course, but it plays a very important role there: a melting smile is reported by many women to be one of the sexiest things they notice about a man.[12] One theory about the origins of smiling is that it evolved to signal trustworthiness.[13] Flashing a "genuine" smile is difficult (though easier for some people than for others), and for nearly all of us, it is much easier to do if we are genuinely feeling relaxed and warmly disposed to the person at whom we are smiling. Furthermore, a genuine smile stimulates warm feelings in others and a willingness to trust the smiler—which would be a dangerous form of susceptibility if the smile were not a genuine signal of trustworthiness.

Smiles Are Signals Too

Together with colleagues from the Toulouse School of Economics in France and the Max Planck Institute of Evolutionary Biology in Plön, Germany, I have recently been trying to test this theory of smiling in the laboratory.[14] We do this using a well-known form of experiment called a trust game. One person is given a sum of money and has to choose whether to keep it or to send it to a second person, whom we can call a trustee. If the money is sent, it is immediately tripled. The trustee then has to choose whether to keep the money or to send some of it back to the original subject, the "first player." You might think this would lead to an obvious outcome: trustees would always keep the money. But they don't always do so. If they did, and if that were what everyone expected, none of the first players would ever send the money, and the experiment would always end boringly at the first stage. It's now well known that some first players do send the money, and trustees return money often enough to justify the optimism of the first players. This outcome has been corroborated with scores of trust experiments in many different settings; in our own version, a little under 40 percent of first subjects sent money, and around 80 percent of trustees returned money.

The novelty in our version of the experiment is that before the game starts, we give our trustees the chance to make a short video clip in which they introduce themselves to the first players. We say nothing to our subjects about smiling, but almost all the trustees try hard to project a convincing smile to their viewers. Their success at doing so turns out to make a difference to how much the first players are willing to trust them by sending the money: those whose smiles are rated by viewers as in the "more genuine" half of the group receive money just over 40 percent of the time, while those who are rated as being in the "less genuine" half of the group receive money around 35 percent of the time.[15] The difference, a little over 5 percentage points, is not huge, but it's big enough over time to make quite a difference to the success of the genuine smilers.[16]

Still, showing that genuine smiling encourages people to send you money doesn't prove that smiling is a signal of trustworthiness. Pointing a pistol at them might have the same effect. So what makes smiling a signal rather than just a way to manipulate the gullible? It turns out that those who smile more convincingly are the ones whom it pays to trust: they really do return more money overall. And there's an interesting twist to this: they return more money not just because they do so more often but also because they have, on average, more money to return (so they have more to smile about). A smile is a reliable signal not just of a person who's more disposed to share their pie with you but also of a person who has a bigger pie to share. This result suggests that the ability to be charmed by a smile was not just some dangerous susceptibility for our ancestors but offered an adaptive advantage.

In our experiments, we've found, too, that the ability to smile convincingly doesn't just come as a package with a more generous disposition, something that either you have or you don't. When subjects play for larger stakes, they put more effort into their smiling, and it shows. (Think what

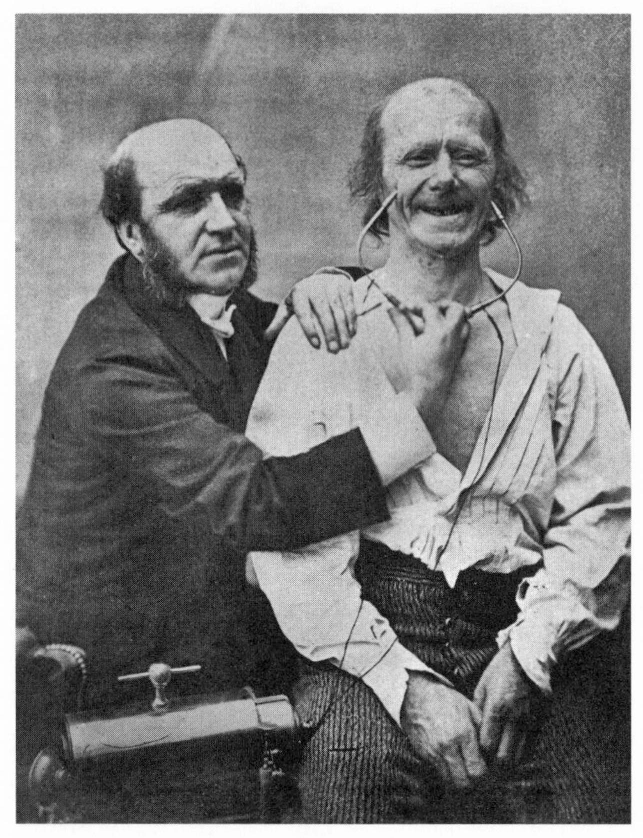

The nineteenth-century French psychologist Guillaume Duchenne using electrical stimulation to induce artificial smiles, thereby demonstrating that "genuine" smiles use more than the muscles around the mouth. © Hulton-Deutsch Collection / Corbis.

happens when you have an important job interview: you put a lot of mental effort into developing the frame of mind in which you can behave charmingly to your interviewer.) In short, it's difficult to smile convincingly if you don't have a reason to do it. As with most difficult activities, people make a bigger effort to smile if they see a larger gain from doing so. And natural selection seems to have rewarded them for that effort, since those who make the effort get better results than those who don't.

Like other kinds of signaling, then, smiling has evolved as a way for people to decide (faster than if they had to work it out explicitly) whom they can afford to trust. Smiling plays an important part in sexual flirtation, but, like so much else about sex, it's also part of a much bigger story about human cooperation at all levels. Cooperation requires choosing partners. Competing against others to be chosen by the most desirable partners stimulates advertising. In a species as cooperative as our own, advertising is everywhere: in every gesture, every movement, every utterance. To succeed, it's essential to stand out somehow, and standing out often requires being different from the background. That helps us to understand the origins of the astonishing variety of human cultural behavior.

A World Saturated with Advertising

Some advertising tries to be attractive to everyone, trumpeting generally valued attributes like reliability or value for money. But often it signals particular qualities that only some people care about: design, taste, or style. This kind of advertising is not necessarily less cost-effective: it concentrates on a smaller audience to make its pitch more intensely. Sexual signaling is no less protean in its purposes: some of it is about advertising qualities that most people would find attractive in a mate: health, wealth, or intelligence, for instance. One theory of the evolution of the peacock's tail suggests that it advertises the superb fitness of its bearer because only a really fit peacock could possibly get away with the ridiculous metabolic expense of growing such an absurd appendage and bearing the insane handicap it imposes on its chances of escaping predators (this is known as the "handicap" principle).[17] As with the Claudia Schiffer car advertisement, or the bread and circuses of the Roman emperors, the expense of the display is—according to this theory—not an incidental cost but the whole point. Of course, it's not the handicap in itself that is attractive: it's the underlying high quality of the signaler, who, by showing that in the presence of the handicap he can do at least as well as this rivals, inspires the belief that the underlying quality more than outweighs the handicap.[18] And the handicap may not have been at the origin of the display's attractiveness: the first peacocks' tails may have been eye-catching enough to appeal to an existing sensitivity to bright colors on the part of females, and it may only have been once they grew substantially that the handicap began to count.

Other kinds of advertising signal "niche" qualities that only some audiences find attractive: while (almost) everyone might prefer an intelligent mate, not everyone prefers an intellectual one. Height in a partner might be a sign of health, but it's attractive only up to a point: taller people usually prefer taller mates, and shorter people prefer shorter ones. So an alternative theory to explain the evolution of the peacock's tail is that it started out not as a signal of greater fitness but as a niche taste. Large, bright tails simply appealed to the kinds of female that preferred large, bright tails.* Once such females were around in sufficient numbers, their taste itself became adaptive because females who shared it would have male offspring that would appeal to other similarly inclined females. This is known as the "sexy son" hypothesis,[19] and it's not incompatible with the handicap principle: peacocks' tails might have arisen from both advantages. There has been independent confirmation of this suggestion from experiments showing that female birds who are initially indifferent to a particular male may become attracted to him suddenly if they observe him apparently surrounded by other admiring females.[20] You can confirm this tendency yourself in the species *Homo sapiens:* in public spaces such as trains, airports, and restaurants, both men and women go to great trouble to signal, through ostentatious use of their smartphones, just how popular they are: they clearly believe this stimulates interest in them rather than discourages it, and they seem afraid of being thought of as losers if their mobile devices are not continually chirping and bleeping.[21] The fact that it's a relatively low-cost strategy means, of course, that it's likely to be heavily overused.

Charles Darwin, who developed the theory of sexual selection in his book *The Descent of Man*, was the first scientist to realize that specialized advertising could explain rapid, divergent evolution in the characteristics of closely related populations.[22] It's striking how different from each other many closely related bird species look, such as different species of bower-

* As this hypothesis indicates, scientists can launch successful careers on the enunciation of tautologies: the trick, obviously, is not to stop there.

birds or birds of paradise. The reason is almost certainly that the parent species clustered into subpopulations according to the tastes of the females who were selecting between types of masculine display. The females with one niche taste bred with one group of males, while those with a different niche taste bred with a different group of males. The resultant reproductive isolation—due to the females' choice, not to geographical separation—led the subpopulations to diverge rapidly into distinct species. What stops *Homo sapiens* from likewise separating into subspecies based on, say, tastes in rock bands is the fortunate fact that most children never inherit their parents' taste in music.

Darwin was also aware that on even shorter timescales, sexual selection might lead to striking variations in superficial characteristics between otherwise very similar subpopulations of the same species. This observation had an important application to human society, where there are conspicuous differences in hair and skin types across geographical regions. Darwin realized that if sexual selection could explain why human populations from different continents could look so different when, as he believed, they were so closely related to each other, there would be an important political as well as a scientific implication. Adrian Desmond and James Moore have argued that Darwin was strongly motivated by his detestation of the institution of slavery, which he had seen at close hand during the voyage of the *Beagle*.[23] He reacted in particular against those defenders of slavery who argued that the black races were a different species from the white races. Darwin hoped that if it could be shown that all the human races were descended from a single human ancestor, the case for mistreating the members of what he called the savage races would be much weaker. In order to do that, he had to explain why the black races and the white races and the brown races looked so different when they were really so similar. Sexual selection came to be crucial to that explanation: the fact that dark people might prefer on average to mate with dark people, tall people with tall people, and so on could lead to rapid divergence in the superficial and visible characteristics of human populations even if their more fundamental characteristics remained very similar.

It's worth emphasizing the importance of Darwin's theory of sexual selection for explaining human diversity, because substantially divergent evolution between closely related populations is surprising unless those populations are physically isolated (like the Galápagos finches on different islands). Not only do most closely related species show strong resemblance, but it is striking how often natural selection has even produced convergent evolution: that is, it has found functionally similar solutions to the same problem on multiple occasions. Thus birds and bats have both evolved wings. Arctic and Antarctic fishes have both evolved a way of stopping their blood from freezing in the cold waters by synthesizing antifreeze proteins, but they do so using different proteins and with quite different genes that encode for them.[24] Anteaters in Australia and South America have developed long snouts, but they are not at all related, and their common ancestor almost certainly didn't have such a snout.[25] The gene for lysozyme, an enzyme that is part of the immune system, has evolved convergently in different cellulose-digesting mammals.[26] Garter snakes and clams have independently evolved a very similar mechanism that gives them resistance to toxins in their prey.[27]

So there are several examples in which natural selection in a reasonably static environment has implemented the same functional solution to a problem in two or more different anatomical ways. Sexual selection, in contrast, is about the coevolution of the characteristics of an individual with those of somebody else, a mate or a rival: as the behavior of the mate or the rival changes, so does the way in which it's best for the individual to respond. Sexual selection is trying to hit a moving target, and two closely related populations can move the target in quite different directions. Again and again we see examples of divergent evolution between closely related species, in spite of their similar habitats and similar histories.

This coevolution is also the source of sexual conflict, which, as we saw in chapter 1, occurs even over when and how to mate: each modification in the strategy of males poses a new challenge to females, and vice versa. It underlines something about the often strikingly wasteful character of sexual selection. There is wasteful advertising: wouldn't it be easier if peacocks found a simpler way to signal their fitness? There is wasteful investment in male strategies to subdue females and female strategies to escape persistent males. There is wasteful competition between rival males, which takes all forms from the battles of dominant sea lions to the slaughter of young men in military campaigns (this doesn't mean that any individual male is behaving irrationally, just that competition makes all collectively worse off). Darwin himself famously wrote, "What a book a devil's chaplain might write about the clumsy, wasteful, blundering, low and horribly cruel work of nature."[28] For people who have seen how admiringly Darwin wrote about the beehive and about some of the perfectly formed organs of the animal body, it can sometimes be a surprise that he was completely aware of the more rebarbative results of evolution by natural selection. But he was as lucid about the horrors of the natural world as he was enchanted by the beauties of its design.

Advertising and Enchantment

Many of the achievements of our civilization started out as ways for men and women to impress each other. Whether you call them wasteful depends on your perspective. The cosmetics industry is in some sense a vast waste of money: if the purpose of cosmetics is signaling, why don't we all just get DNA tests to signal our genetic qualities directly? The answer is that even if signaling may be the reason why cosmetics increase someone's sex appeal, natural selection has shaped us to respond to the signal and not to the underlying reason. Characteristics that seem sexy in a potential partner don't become less sexy when we reflect that they originally attracted our ancestors because they signaled something else, such as fertility. The knowledge that someone is vasectomized or on the pill need not diminish their sex appeal one bit. Conversely, a certificate of your DNA profile will not evolve to look sexy on less than a geological timescale, which is longer than most people are prepared to wait to find a sexual partner.

Competitive sports are also a form of signaling. Inciting young men to compete to kick footballs through goalposts sounds like the most fatuously wasteful occupation ever invented, yet the delight it brings to the millions of spectators around the world suggests that "waste" is never quite what it seems. And this is as true of "high" culture as of the mass-market kind; poetry and song both started out as instruments of display, whatever elaborate reasons we may give for their effect on us now. When Shakespeare writes that "love is not love / Which alters when it alteration finds," his readers should not be fooled. He may be writing a sonnet, but he is not engaging in careful factual description of people's feelings. He is not commending a psychological hypothesis—certainly not one so overwhelmingly at odds with the evidence accumulated over centuries. He is not developing an objective thesis on the human condition out of a wish to advance our knowledge or the search for the Platonic form of beauty. He (or at least his narrator) is signaling: pleading to be taken seriously, pleading all the more urgently because deep down he has doubts about his own constancy too.[29]

The enchantment woven by love is no less real for being elusive and ineffable: its elusive and ineffable nature serves powerful biological needs precisely because it bypasses the explicit reasoning of the calculating brain. Indeed, the reluctance we sometimes feel to look too closely at the hidden sources of that enchantment is exactly what you'd expect a sophisticated process of advertising to encourage. Manufacturers of perfumes never show you the pipes and pumps in their factories, and most upmarket restaurants would never allow the diners anywhere near the kitchen. I once saw a sign on a small furniture store in east London that read: "We buy junk and sell antiques."* It was funny not because it was false but because it was a truth other stores would never acknowledge, just as in politics, according to the journalist Michael Kinsley, "a gaffe is when someone accidentally tells the truth."[30] The vendors of some of the most potent sexual attractors of all, cut diamonds, go to enormous lengths to disguise the fact that they are basically buyers of pebbles and sellers of dreams. The alchemy by which the former are transformed into the latter is less charming the more you know about it. As their customers, we don't necessarily want to know any better: if an engaged couple really believes that the exchange of one type of pebble gives their relationship more staying power than another type of pebble, only a very heartless observer would want to disabuse them. "Know thyself," said the inscription in the temple of Apollo at Delphi, but it is a counsel of perfection, and perfection can be the enemy of the good. If you've been given

* There's even a website called Webuyuglyhouses.com. In case you wonder what they sell, the answer seems to be "luvly" houses.

the placebo in the randomized double-blind trial of a drug that could save your life, you'd rather never know.

If we sometimes have good reasons for not trying too hard to untangle the deceptions practiced on us by others, at other times even active self-deception may be in our interests. The biologist Robert Trivers has suggested that self-deception may be a natural by-product of features of brain organization that help us deceive others. It's easier to hide from others evidence that is inconsistent with what you say if you also keep that evidence out of your own conscious mind, whose processes are more visible to the outside world. Some recent experimental evidence supports this theory, suggesting that various self-serving biases in people's beliefs tend to weaken when their attention is given a heavy cognitive load to process.[31] This finding pushes to its logical conclusion a quite reasonable preference we all have at times for what might be called minimal communication under stress: sometimes you'd rather have an uncomfortable discussion over the telephone or by e-mail rather than in person, because it reduces the effort you have to invest in mastering your feelings and trying not to reveal too much. Self-deception is the most minimal communication of all, because you hide from your own conscious awareness the things it would be costly to reveal to others, things it would cost you continued effort to suppress if you acknowledged them.[32]

Indeed, self-deception flourishes in precisely the areas of life where these costs matter most: in business, in politics, and in love. The most effective salespeople are often those who really love their product, whatever natural tastes they may have had to suppress to reach this state. Many politicians, whose every utterance is scrutinized for the consistency of its logic and body language, find it easier to persuade themselves of the truth of various conveniently self-serving beliefs before they start to promote them to others. That's why it may not always make sense to ask whether a politician is lying or is sincere: the answer may be "Both." The nineteenth-century British prime minister Benjamin Disraeli memorably captured the attractions of this strategy when he described his opponent William Gladstone as "a sophistical rhetorician, inebriated with the exuberance of his own verbosity." Disraeli's pairing of the characteristics of sophistry and inebriation is at first surprising—you'd expect the latter to make the former impossible—but as the logic of self-deception has shown, it's absolutely accurate. And any lover will be more ardently convincing if he has already silenced whatever internal doubts he may have about the extravagance of his declarations. Before deciding whether to propose marriage to someone, you should read everything you can find about divorce statistics, but once you've decided to go ahead and make the proposal, you should forget everything you've read.

In short, there are good reasons for our reluctance to dwell too closely on the underlying biology of our emotional commitments: natural selection has made us that way because the commitments are more effective when we and our partners lack the lucidity to unravel their origins.* But these reasons, persuasive as they are, don't make the biology go away. Love can weave enchantment that, like all magic, fosters in the enchanted the illusion that the present moment is inexplicable and ineffable, beyond all purpose outside itself. To a person in love, the irreplaceability of the loved one is unarguable and its origins in our endocrine system an irrelevance. To those whose lovers have left them, the memory of abandonment, even years later, can be evoked by the cadence of a forgotten voice, by a half-caught aroma of jasmine or by a whisper of silk. Even so insubstantial a reminder can leave them emotionally sandbagged, in a state that language seems wholly inadequate to describe.[33] We often don't even want to understand why we feel that way: we fear that understanding would leave us forever unable to succumb to enchantment again.

Nevertheless, once we ask ourselves the question, we should be lucid about the answer: if we are susceptible to enchantment, natural selection has had its reasons for making us so. When John Donne writes of the passion that makes "one little room an everywhere," or Jacques Brel sings of a love that drapes in beauty the ugliness of the suburbs, inviting the enchanted to admire, it is because the act of doing so makes the enchanted stand still. If you can make a man think one room is everywhere—especially a man with

* Scientific hypotheses can be enchanting too: as Fernando Pessoa put it, "Newton's binomial is as beautiful as the Venus de Milo, but fewer people realize it" (Pessoa 1987, 238, my translation).

as much testosterone as John Donne—you dampen his urge to wander elsewhere. The very lives of your children may depend upon it.

Both the magic and the menace of sexual love have become so elaborate in the human species because cooperation is central to our social life, and because making children and preparing them for life demands a degree of long-term commitment surpassing anything else seen in nature. It's a commitment that sets up unprecedented tensions between the long-term and the short-term interests of the partners and complicates dramatically the nature of the signaling between them. In chapter 4 we look at why and how our ancestors came to develop that particular model of social behavior, which took us in a very different direction from the other social primates who are our closest natural cousins. It meant that our capacity for sustained teamwork had a much larger influence on our survival and reproduction than it has done for any other mammals (social insects such as ants, bees, and termites are great cooperators too, but males play a much smaller part in the team, and these insects' sex life has little to make us envy them).

You might think that our massive dependence on sustained teamwork would have led to strong selective pressures to make us nicer or more dependable. There's an element of truth in that: when successful and sustained teamwork matters, natural selection may favor qualities such as generosity and loyalty. But it may also, and simultaneously, favor the talent for manipulating others, for benefiting from the contribution of others while minimizing your own. Above all, it has favored the talent for subjecting the behavior of others to the sharpest and the most unforgiving scrutiny. Enchantment and suspicion have entwined themselves together around the solid trunk of human social life.

The Birth of Suspicion

Suspicion is born out of the richness of the signals that human sexual partners send to each other and the ample opportunities they create for confusion and manipulation.[34] Their behavior is a continuation in more elaborate form of a simple interaction that is everywhere in nature. To understand it, we need to examine what it shares with the behavior of other species as well as what makes it different. Vast and various though it may appear in the natural world, the range of strategies used by males to gain access to females amounts in the end to a series of elaborate variations on only two themes: defeating rival males through force or cunning, so that the victor has the choice of females to himself, and impressing the females so they choose him over rival males. If the victor has the choice of spoils, that's a setting that favors flamboyance and risk taking among the males. Aristocratic strategies will always thrust aside the prudent and bourgeois virtues: being good enough is never quite good enough. In this world, suspicion is a guy thing: males are each other's rivals and the objects of their most intense mutual scrutiny; females largely take what they can get. But even when persuasion rather than force is the weapon of choice, the only thing that impresses the females, in many of the species where they get to choose, is the quality of a male's sperm, which is all he will ever contribute to the partnership. However elaborate the signaling, it is always aimed at communicating that one attribute.

Given how cheap sperm is to produce, and how many females the most successful male can inseminate, this criterion of female choice creates a strong incentive to aim at being the very best; there's no reason why females should settle for second best, any more than music lovers should settle for listening to a second-best recording when the best is also available and can be easily reproduced. They have a strong incentive to scrutinize and select the males, but once the deal is done, there's no need for further vigilance.

But man cannot live by sperm alone, and nor can a good number of other species. A number of males offer food or protection along with their sperm. In some species—in the dance fly, for instance, and even more spectacularly in those species of cannibalistic spiders we met in chapter 1—the food and the sperm are delivered together. The male has an obvious interest in making sure the link between the two offerings is unbreakable: from natural selection's point of view, the investment of food is worth making only if it really is his sperm that fertilizes the female. In others—in many birds, for instance, as well as in human beings—the food and the sperm are not delivered simultaneously but can arrive in a series of interleaved deliveries over quite a long period. Because the female now has an interest in ensuring that these deliveries are really made as advertised, suspicion about the quality of the male's promises is now a natural and adaptive psychological trait. But suspicion goes both ways: the male's interest in ensuring that only his sperm gets to fertilize the female in whom he is investing his food remains undiminished, and he can now no longer rely on simultaneous delivery to ensure that. This isn't just a matter of accident: natural selection sets the females' interests subtly in opposition to his own. Since the emergence of pair bonding, it has occurred to females of a number of species to wonder whether the two things—food and sperm—absolutely have to come as a package deal. Perhaps the food could come from one male and the sperm, or at least some of the sperm, from another.

To put it more accurately, it is natural selection that, by transmitting the restlessness in a woman's loins to her daughters and her granddaughters, has been doing the wondering for all of them. That restlessness first made itself felt many millions of years ago, but already the countdown to *Anna Karenina* had begun.*

* Novels are notoriously unreliable as a source of evidence, since novelists are fascinated by the exceptional, but they can aptly express certain ideas that are amply corroborated from other sources. Stendhal, writing in 1830, has a sardonic description in which Julien Sorel imagines himself loving a sophisticated woman in Paris: "He loved with passion, and he was loved. If he left her for a few moments, it would be to cover himself with glory and to deserve her love all the more." But Stendhal adds that any young man who had actually been "brought up among the sad realities of Parisian society would have been woken at this point in his novel by a cold irony, and his grand actions would have disappeared, along with any hope of achieving them, to be replaced by the well-known maxim: leave your mistress's side and you'll be cuckolded—alas—two or three times a day" (Stendhal 1962, 68–69). A century later, Joseph Keller wrote of his novel *Belle de Jour* (published in 1928 and later made by Luis Buñuel into a film starring Catherine Deneuve) that "what I tried to do was to show the terrible divorce between the heart and the flesh, between a true, vast and tender love and the implacable demands of the senses. With a few rare exceptions, every man and woman who loves for a long time bears this conflict within. It may be felt or not, it may slumber or it may tear, but it exists" (Keller 1928, 10, my translation).

FOUR

Social Primates

In his country,
spotted crabs
born in their mother's death
grow up with crocodiles
that devour their young.
Why is he here now?
And why does he
take those women,
a jangle of gold bangles
as they make love,
only to leave them?

—"What She Said," from *Poems of Love and War,* translated by A. K. Ramanujan (in Ramanujan 1985)

Competition and Cooperation

Charles Darwin is widely thought to have bequeathed to us a vision of human society that, for all its conventional pieties about the importance of cooperation, is pitilessly competitive to its very core. "A war of all against all" is a phrase commonly quoted from Thomas Hobbes but often considered an appropriate description of Darwin's own social vision, particularly by those who know *The Origin of Species* but not Darwin's later works.[1] In the closing paragraph of *The Origin*, Darwin wrote: "Thus, from the war of nature, from famine and death, the most exalted object which we are capable of conceiving, namely, the production of the higher animals, directly follows."[2]

Despite the superficial similarity of the two concepts, Darwin's vision of the war of nature was utterly different from that of Hobbes. It was a war that, far from pitting all against all, needed and rewarded teamwork. The observant Darwin could hardly have failed to notice what any reader of this book can easily verify hundreds of times a day: human society surrounds us with people on whose cooperation we depend. From the moment you get up in the morning, you depend on other people to help make it through the day, whether it's the suppliers of your toothpaste and your morning coffee, the person who fixed the water leak you never knew about in the middle of the night so you could have your morning shower undisturbed, the driver of the bus you take to work, the journalist whose story made you giggle in the train, or the colleagues in your office, shop, or factory. In a true war of all against all, you would hope that all of these people had faltered, made mistakes, had in some way failed to do their jobs properly: that's what it means for someone to be your competitor. In fact, as a moment's reflection attests, we depend so much on others to do their jobs well that we hardly notice the smoothness of the social fabric they have made for us; the occasional incompetent stands out all the more annoyingly because most of the time the process works so well. It's a tribute to the effectiveness and omnipresence of cooperation that we can so easily forget that it happens at all.

Still, if human society is as cooperative as I have just made it sound, why does it all feel so stressful? A phrase like "the smoothness of the social fabric" hardly does justice to the tension almost everyone in a modern society feels as they juggle the demands of friends, family, and colleagues. It barely captures the pressure of other people's expectations, the anxiety as to whether we shall be able to deliver what other people want from us, whether we shall ever be considered up to the mark. You might be forgiven for thinking that human society is like one of those "good news, bad news" jokes they tell about doctors: "Congratulations, *Homo sapiens*, the good news is that you've evolved to be an extraordinarily cooperative species. The bad news is that cooperation is just as awful an experience as competition."

Certainly cooperation is far from the cuddly ideal of the hippie generation: at the very least it involves setting higher standards for our behavior toward others than our natural inclinations recommend and eternal vigilance about whether our own and others' behavior really meets those standards. But a deeper explanation for all that stress is that cooperation rarely happens at the level of society as a whole: it takes place among groups within that society, groups whose membership can be highly fluid. Some of those groups are in competition with each other, and individuals may compete with one another to be accepted in the most influential and successful groups. It makes a big difference to your well-being which groups you can join, and their acceptance or rejection of you is a conspicuous and stress-inducing event. None of this will be surprising to anyone who has a clear memory of their school playground. None of it would be surprising to a primatologist, since we human beings are a species of social primates, and social primates are nature's experts at building (and breaking) coalitions. And none of it would have surprised Charles Darwin, who was acutely aware of the primate origins of our nature, even though he had drawn a veil over these in *The Origin of Species.* In fact Darwin had the remarkable insight that natural selection might positively encourage altruistic qualities because they could help coalitions to be successful in competition against rival coalitions. Here is a passage from *The Descent of Man*, Darwin's first serious discussion of how natural selection shaped human societies, published more than a decade after *The Origin of Species:*

> When two tribes of primeval man, living in the same country, came into competition, if (other circumstances being equal) the one tribe included a great number of courageous, sympathetic and faithful members, who were always ready to warn each other of danger, to aid and defend each other, this tribe would succeed better and conquer the other. Selfish and contentious people will not cohere and without coherence nothing can be effected. A tribe rich in the above qualities would spread and be victorious over other tribes . . . thus the social and moral qualities would tend slowly to advance and be diffused throughout the world.[3]

It's clear that the qualities that Darwin thought human beings needed to survive in the "war of nature" were very different from those that later generations would associate with the term *Darwinian*. Our primate nature is the key to understanding this, because life as a social primate is all about managing cooperation in a world of shifting coalitions. This is very far from Hobbes's "war of all against all," but it's also a long way from being a serene parade of agreeableness. Certainly, the determinants of fitness include our ability to cooperate with others in our group, the social intelligence to work out who else can be trusted to cooperate with us, and the ability to persuade other members of our group to work with us and not against us.[4] But competition is present at many levels too. There's competition between individuals over basic access to economic resources: someone may be collaborating with you on an important project and still be trying to get a bigger share of the benefits than you get. Competition also takes place in other dimensions, notably between coalitions of individuals, and between individuals over the prize of access to powerful coalitions. The tension between those forms of competition is at the heart of primate social life. At any moment you are seeking to cooperate with some of the other members of your group, but at the same time you are intensely anxious about which groups you may be allowed to join (the groups are obviously more complex in human societies). Your fitness is going to depend not just on what you do but on what you can induce the other members of your group to do with you and for you, and on which groups you can persuade to count you as one of their own.

This is a social environment perfectly suited to inducing stress. In dominance hierarchies in most primate societies, low status is associated with characteristics that are usually good predictors of stress, such as a lack of autonomy and social control and a high degree of unpredictability in outcomes. Studies of human societies show similar results. There's a popular belief that stress goes with high-status jobs, but the truth is very different. A longitudinal study of British civil servants by Michael Marmot and his colleagues shows that stress-related illnesses such as cardiac disease show a strong negative correlation with rank.[5] The lower down the hierarchy you are, the more likely you are to have somebody pushing you around. This is a very stressful situation, and it is associated over time with significantly greater risk of stress-related illnesses.

It was believed until recently that in nonhuman primate groups, exactly the same thing was true, but when it was first possible to measure cortisol levels (a physiological indicator of stress) among primate groups in the wild, some surprising results emerged. An interesting paper by Muller and Wrangham showed that cortisol levels were *positively* correlated with dominance among wild chimpanzees.[6] The authors suggest that this is because dominant chimpanzees spend a great deal of energy maintaining their dominance—more, certainly, than senior civil servants do. It's a tough and uncertain life for chimpanzees at the top, and their cortisol levels are elevated because of the continual uncertainty about how long they're going to stay on top and the continual efforts they have to make to ensure they stay there. Frans de Waal's book *Chimpanzee Politics* describes such a life in gripping detail.[7] Nor are the findings confined to chimpanzees: high levels of stress hormones have been reported for alpha male baboons.[8]

So primate existence is pretty stressful either way. You get stress if you win and stress if you lose. The explanation seems to be precisely that small differences in behavior may result in very big differences in outcomes: two individuals who are initially alike may have dramatically different life trajectories if one is accepted as a member of powerful coalitions while the other is not. (This may be part of the appeal of social structures such as caste systems.) It's not surprising, therefore, that even though we're surrounded by potential collaborators, we live in continual fear that our potential collaborators will decide to collaborate not with us but with somebody else.

Primate Negotiations

In our own species, unusually among social primates, a lot of the teamwork is undertaken by men and women together. There's a lot to gain from that cooperation, but the gains are often unequally shared. Why should that be? The answer lies in powerful economic forces that have shaped the relative bargaining power of men and women over the course of our evolution. One of the most robust findings in economics is that it pays to be indispensable. The only plumber available to fix your leak on a Sunday evening may be able to charge you an outrageous sum for the privilege. But if many plumbers happen to be in town for a convention, you can expect to get a better deal: competition drives down prices. What counts is not the indispensability of the service in general (fixing the leak) but of the particular plumber who does the job. Similarly, if you need a new set of keys to get into your house, the keys may be indispensable to you, but the particular locksmith who supplies them to you will not be. It is thanks to competition among locksmiths that you don't end up having to remortgage your house every time you lose your keys.

In general, then, the more fiercely you have to compete for the favors of the people with whom you hope to cooperate, and the fewer rivals they face for the privilege of collaborating with you, the less likely you are to be able to bargain for a large share of the benefits of that cooperation. We can therefore chart the way in which the gains from teamwork have been distributed between men and women by looking at how the intensity of competition among each sex for the favors of the other has changed during our evolution. In a phrase, men used to be the dispensable sex. The fact of sexual reproduction meant that the service they provided (sperm supply) was indispensable—but, like the locksmiths, the individual suppliers were not. However, over the course of the development of the human species, they managed to become progressively more indispensable, and they exacted a price from women as a result. That price is not inevitable: many features of modern society have made men more dispensable again. But we shall not understand the terms of the gender bargain until we understand the way competition shapes it, in other species as well as in our own.

Competition isn't just a matter of how much rivalry exists between males for the favors of females, as compared to the rivalry that exists between females for the favors of males: other factors make a difference, including how much the parties are committed to each other over the longer term. But that rivalry is a good place to start. Male dance flies have found a way to compete for the favors of females by offering food, and this has made the females compete in turn for access to this scarce resource. Human males have managed to do something rather similar: in a nutshell, they learned to contribute substantial parental care, and human females began to compete with each other to find and hold on to the most reliable providers of that care.

To see how that has happened, let's begin by noting what we share with other social primates, in particular with our closest cousins, the great apes: orangutans, gorillas, chimpanzees, and bonobos. Great apes differ from most other primate species in that their societies are patrilocal. This means that adolescent females leave the communities where they were born to join other communities. Females are wanderers, and in their wanderings they exercise some degree of choice over their mates. From that point on, there are significant differences in the different species' living arrangements. Orangutans remain famously solitary, whereas gorillas live in harems controlled by a single dominant male. He is much larger than the females as a result of the winner-takes-all rivalry among males, a rivalry whose reward is control of the entire harem. However, chimpanzees and bonobos, the closest to us of all the great apes, have a very different arrangement. Females of these two species live in small groups and mate promiscuously with many males. As you'd expect, because they don't have to compete for control of a harem, males of these species are relatively smaller than male gorillas. To be more accurate, they're smaller in almost all dimensions except one. They have massively larger testicles—nearly four times as large as the gorilla's, although their body weight is little more than a quarter of the gorilla's. As Roger Short, Alexander Harcourt and others have shown, large testicle size is closely associated with mating systems in which females mate with multiple males.[9] This is for the simple reason that if males have no realistic hope of monopolizing sexual access to females, it is in the reproductive interest of each male to ensure that his sperm is as abundant as possible in order to increase the probability that it fertilizes the females in preference to that of rival males (this process is called "sperm competition").[10] There are important differences between chimpanzees and bonobos: in particular, bonobo females spend more time in stable groups than do chimp females, and their cooperation gives them more power over males.[11] But the multiple matings in which each female engages are a common and striking feature of their social life.

Diverse as these mating arrangements may seem, they all involve females' exercising substantial choice. As the primatologist Craig Stanford has put it, "Among the four great apes, males try to control females . . . but females are difficult to control because they are following their own reproductive agenda."[12] Female gorillas give up some of that control of their reproductive agenda once they join the dominant male's harem, but chimp and bonobo females retain it both before and after they move out from the troop of their birth to join another troop, with bonobos in particular reinforcing that control by elaborate cooperation among the females in their troop. What makes it possible for female great apes to follow their own reproductive agenda is a relatively simple fact: once a female has been fertilized, she no longer needs the males, and certainly she has no particular need of the male that happens to have fathered her offspring.

This picture—of great ape females handling with aplomb a lot of hassle from males and maintaining their capacity for autonomy and choice throughout the process—sounds very different from the lot of most human females since the dawn of recorded history. Indeed, since written records began, and until very recently in most parts of the world, women were under the control of their fathers until that control passed to their husbands on their marriage, usually without their having any say in the matter.[13] How and why did that arrangement develop? Why have human females been so comprehensively constrained by comparison with their great-ape cousins? Without written records it is difficult to be sure of the answer, but social behavior leaves records in other ways.

The Gender Bargain among Hunter-Gatherers

One thing we can be reasonably sure of is that the subordination of women must have happened very recently in evolutionary time. The evidence comes partly from the observation of hunter-gatherer societies in modern times, in which women have played a central part in foraging and therefore have exercised a substantial degree of autonomy. Agricultural societies have often been able to impose heavy constraints on women, either by ensuring that they do not work outside the home or by making them do routine work under close supervision in the fields. But foraging societies cannot make use of the labor of women who cannot move, cannot take decisions, and cannot think independently.

Still, the foraging societies that have been observed by modern scholars are few and highly untypical of the societies in which *Homo sapiens* evolved. While it seems a reasonable guess that human females can never have been as subordinate to males in hunter-gatherer communities as they became in agricultural societies, it would be good to have corroborating evidence. Two other kinds of evidence can be found directly in the human body, because one of the marvels of natural selection is that past behavior leaves its mark on current anatomy.

The first kind of evidence comes from our brains and from their cognitive capacities as psychologists have been able to measure them. As I discuss more fully in chapter 5, the average cognitive capacities of men and women differ in a number of respects. Men score significantly higher on average than women in some tests of spatial reasoning, such as tests of the ability to imagine whether the object in one picture is a spatial rotation of the object in another picture. Other cognitive tests, such as exercises in verbal reasoning or tests matching emotional states to faces in photographs, yield significantly higher scores on average for women than for men.[14] There are also some well-documented differences in preferences between men and women, such as the often-replicated finding that women are on average more risk-averse than men.[15] It's possible to debate whether such results are affected by the way in which the tasks are presented to the subjects, and there is also room for reasonable disagreement as to whether the dimensions in which women are stronger are more or less important than those in which men are stronger. But these qualifications are irrelevant in this context. Cognitive differences on average between men and women are very important for my argument here, but that is because of how small they are, not how large they are.

Because claims about gender differences in intelligence, like claims about racial differences in intelligence, have caused such acrimonious political debates in recent decades, many people who are averse to being considered racist or sexist have avoided trying to find out about them or even to think very much about them at all. This means that most people have never noticed something about gender and racial differences that was evident to Charles Darwin and that ought to become evident once again to us today. According to Darwin's view of sexual selection, the likelihood of finding gender differences is higher than the likelihood of finding racial differences. More exactly, this view implies that we should not expect to find important inherited differences in cognitive abilities between different human populations (including between races). Thus it's neither very surprising nor very informative when it proves difficult to find hard evidence for such inherited differences (raw differences reflecting environmental factors are relatively easy to find).[16] Sexual selection, by contrast, should lead us to expect much bigger differences between men and women than in fact there are. So our failure to find larger differences tells us something very interesting about the conditions under which our species evolved.

Darwin spent many pages of *The Descent of Man* pointing out that sexual selection works in the first place on relatively superficial characteristics, such as skin color and type of facial hair. This meant that over quite a short period (by the standards of evolutionary time), two groups within a population (say the darker-skinned and the lighter-skinned) could become reproductively isolated even though they lived in the same territory. They would still be capable of interbreeding very easily; they just wouldn't usually want to. Their other traits might begin to diverge through the process known as genetic drift, but that would take much longer. And it would be less likely to happen in the case of traits (such as the various components of intelligence) that were subject to continuous selective pressure from the environment in both populations. Darwin was convinced, as we saw in chapter 3, that under the skin, human beings were very much alike. Indeed, his book *The Expression of the Emotions in Man and Animals*, published soon after *The Descent of Man*, underlined this point with evidence that facial gestures expressed emotions in a remarkably similar way across many different races and cultures.[17]

Modern DNA evidence, showing that human beings are descended from a common maternal ancestor who lived around 140,000 years ago and a common paternal ancestor who lived around half as long ago, reinforces that view.[18] True, there are some important genetic changes that can become established in human populations in shorter periods than that, and the migration of populations out of Africa and across the globe would have produced some differences in selective pressure for such characteristics as height, skin color, and disease resistance. But it's most unlikely that the selective pressures on talent and ability would have produced big divergences between human populations on a comparatively short timescale.[19] This wouldn't happen through sexual selection: since talent and ability are both less visible at first glance and more universally appealing when they can be discerned, their sexual attractiveness should not vary much from one human population to another. And it wouldn't happen through natural selection under environmental pressure either: the importance of talent and ability for general survival would have been very similar in Africa and in other regions of the world. So when modern researchers have difficulty finding hard evidence for large inherited differences in cognitive talent between populations, nobody should be very surprised.[20] There have not been sufficiently important differences in selective pressures between populations since those populations became reproductively separated.

That situation contrasts strikingly with selective pressures affecting gender differences. Given the logic of sexual selection, we ought to be very surprised at how hard it is to find significant differences in overall cognitive abilities between men and women. There's no reason for the characteristics women have needed to appeal to men to be anything like the characteristics men have needed to appeal to women. In many other species, males and females are utterly different (think of all those gorgeously colored male birds). They differ in appearance, in behavior, and even in size (the male gorilla is twice the weight of the female). It's true that sex differences are less pronounced among mammals than among birds, for instance, but important differences remain. Men are 15–20 percent larger on average than women, and their upper body strength is substantially greater. That's because muscle tissue is expensive to grow and maintain, and natural selection has sensibly grown more of it among the males who needed it more in the conditions of our hunter-gatherer existence. Brain tissue is far more expensive than muscle to grow and maintain, yet natural selection has nevertheless given equally sophisticated brains to men and to women.

It's just possible that the mutations never arose that would have enabled the processes of embryonic and child development to produce less sophisticated brains in one than in the other (though there are qualitative differences between men's and women's brains that certainly have a genetic basis).[21] But there's another compelling and much more plausible explanation: men and women developed equally sophisticated brains because both faced equally sophisticated cognitive challenges throughout almost all of human evolution. On this view, the subordinate and dependent condition of women that has characterized relatively recent centuries cannot have obtained for most of the time since we diverged from the chimpanzees and bonobos.

A second piece of anatomical evidence suggests that women in most hunter-gatherer societies were a good deal more autonomous than they came to be later, after the adoption of agriculture. This is the fact that the testicles of human males are of intermediate size between those of gorillas and those of chimps. This strongly suggests that human females were mating fairly often with more than one male during a single estrus cycle, albeit not nearly as often as among chimpanzees. They did so often enough, in fact, for it to be adaptive for males to engage in sperm competition, as chimpanzee males do. And this conclusion in turn makes it likely that women exercised a significant degree of choice over their mates.

In addition to testicle size, there is also evidence from penis size, where human anatomy is an even more striking exception to that of the other apes. Men have penises nearly twice as large as chimps and four to five times as large as gorillas do. (They also produce more seminal volume per ejaculate than either chimps or gorillas.) We don't know whether the explanation is sperm competition—the longer penis offers an advantage by depositing the sperm nearer to the uterus—or the greater stimulation that a large penis can provide and for which women might have had a direct preference. But it seems safe to assume that women were likely enough to be mating with more than one man that there was some adaptive value to the difference.

That assumption in turn suggests that women were unlikely to have been entirely under the control of any one man. As we've seen, competition on one side of the relationship increases the bargaining power of the person on the other side.

There are many things we don't know about human mating during the six million years or so that separate us from our common ancestor with chimps and bonobos. We don't know, for instance, whether that common ancestor had a sex life that was more like ours, like that of chimps, or like that of modern gorillas: testicles don't fossilize. So we don't know whether the autonomy of human females was declining in the later hunter-gatherer period or whether it remained relatively stable, only to decline precipitously once agriculture arrived. In many other respects, hunter-gatherer societies changed radically during prehistory.[22] The reduction of size asymmetry between men and women since the earliest species of hominins suggests that force was becoming less important than persuasion in relations between men and women from the earliest times. But at the same time, cooperation among men was becoming more sophisticated, and this development might have been used to women's disadvantage, so a lot of uncertainty remains about what those relations were like. Nor do we know much about the context in which multiple mating by females took place: was it open and acknowledged or furtive and secret?

Naturally, the lack of data hasn't discouraged speculation. The recent book *Sex at Dawn* by Christopher Ryan and Cacilda Jetha has claimed that "our hominid ancestors have spent almost all of the past few million years or so in small, intimate bands in which adults had several sexual relationships at any given time. This approach to sexuality probably persisted until the rise of agriculture and private property no more than ten thousand years ago. In addition to voluminous scientific evidence, many explorers, missionaries and anthropologists support this view, having penned accounts rich with tales of orgiastic rituals, unflinching mate sharing, and an open sexuality unencumbered by guilt or shame."[23]

The evidence these authors cite for the existence of multiple female partners includes the fairly convincing anatomical evidence we have already noted, along with various kinds of evidence from behavior, including women's greater capacity than men for delayed orgasm as well as for multiple orgasms, and the common tendency for female orgasm to be quite noisy, which suggests that natural selection was not selecting strongly for discretion. But the evidence that this was "an open sexuality unencumbered by guilt or shame" is more difficult to interpret than they acknowledge. The anthropological evidence they cite is both conflicting and controversial. For instance, Margaret Mead's famous work on Samoa, claiming evidence of uncomplicated multiple sexuality among Samoan adolescents, has not survived the scrutiny of later ethnographers, and it seems that anthropologists (of all persuasions) have found it difficult to avoid reading into their evidence many of their own fears, hopes, and fantasies.[24] Even solid evidence that such open sexuality has *sometimes* existed would not be the same as evidence that this was the general condition of mankind before agriculture. More important, though, Ryan and Jetha appear to consider that any social practices that persisted for a long period in prehistory would have to be ones that were generally accepted and free of conflict. For instance, they object to a claim by Matt Ridley that monogamy evolved very early among our ancestors, so that "long pair-bonds shackled each ape-man to its mate for much of its reproductive life." Ryan and Jetha comment in response: "Four million years is an awful lot of monogamy. Shouldn't these 'shackles' be more comfortable by now?"[25]

As we've seen, this common tendency to see physical or behavioral traits that have evolved over a very long time as necessarily "optimal" involves a serious misunderstanding of how evolution works. The interests of different individuals can and often do conflict (and even the interests of different genes within an individual can conflict). Predators and prey can evolve traits that are self-defeating; male and female scorpions and bedbugs can have mating strategies of gruesome unpleasantness, as we saw in chapter 1. Men and women can signal to each other in spectacularly wasteful ways, as we saw in chapter 2. Human sexuality could well have remained for four million years or more in a stable equilibrium in which the "shackles," as Ridley calls them, continued to be very uncomfortable for both men and women.

Whether prehistoric sex was like this, or whether it was a more relaxed and open affair, is something about which the evidence is simply too scant and conflicting to allow us to be sure. The duplicity and inconsistency that so often characterize our sexual behavior could be bugs in a system that once worked more easily and openly, or they could be talents favored by natural selection for coping with irreducible conflicts of interest.

The anthropologist Sarah Blaffer Hrdy has suggested a more subtle view than either of these, namely that polyandrous liaisons are likely to have been widespread in hunter-gatherer societies while also being a significant source of tension and conflict—conflict that certain cultural practices could nevertheless help to channel: "Given what a powerful emotion sexual jealousy is, polyandrous liaisons are a risky strategy, dangerous for all concerned. But widely held beliefs about 'partible paternity' help ease some of this tension. In these cultures, semen from every man a woman has sex with in the months before her infant is born supposedly contributes to the growth of her fetus, resulting in chimeralike composite young sired by multiple men. Each possible father is subsequently expected to offer gifts of food to the pregnant woman and to help provide for the resulting child."[26]

Not surprisingly, significant uncertainty about the paternity of a woman's children can have conflicting effects on the incentives of the fathers to contribute to the children's upkeep, as well as cause tension between the different men and between each of them and the mother.[27] Each additional possible father brings into the tent another person who might contribute to child rearing (modern Western societies still have a vestige of this idea in the institution of godparenting). At the same time each addition makes the tent more crowded, potentially diluting the incentives of those already inside to make a contribution. Just where the trade-off has been made by different species, different societies, and different individuals within them is a question on which the evidence is scant and often hard to interpret.

I mention in chapter 1 that many socially monogamous birds are not sexually monogamous, and this naturally prompts the question of whether the extent of multiple mating makes a difference to how well the system of social monogamy works. Researchers who have compared levels of multiple female mating across socially monogamous bird species have observed a significant tendency for greater monogamy to be associated with a system known as "cooperative breeding," in which the female receives help in rearing her chicks not only from her partner but also from other birds (either the partner's siblings or the offspring from previous broods).[28] But there are many exceptions to this tendency, including relatively monogamous seabirds like the puffin that do not breed cooperatively and birds like the fairy wren that have some of the highest levels of multiple mating ever recorded but do breed cooperatively. And the comparison excludes openly polyandrous species such as the dunnock. In fact, it makes comparisons only among the species that practice social monogamy. So the study looks only at the effect of a female's multiple mating on those males whose incentives to contribute to raising offspring are likely to be diminished (namely the female's mate and his siblings) and excludes those whose incentives are increased (namely the other potential fathers). It may tell us that cooperation among potential fathers is reduced in socially monogamous species, but it can't tell us about the comparison between socially monogamous and polygamous ones. This absence of information did not stop the authors from claiming that "the evolution of cooperative behavior is favored by low levels of promiscuity" (they should have added "among socially monogamous species"). This example just shows how hard it is to draw the right conclusions even from a rigorous and impressive scientific study.

There remains a lot of uncertainty both about the degree of multiple sexuality during prehistory and about how openly women controlled their own sexual lives. For the moment, what matters is that our female ancestors almost certainly enjoyed more bargaining power in relation to men than they were ever to do once agriculture arrived. The fact that men were competing for their favors is an important part of the evidence for that claim.

The Changing Needs of Human Babies

Even under conditions of hunting and gathering, human females were more dependent on men than chimpanzee or bonobo females are, and this dependence almost certainly weakened their bargaining power long before agriculture was even imaginable. The reason for this is simply the needs of human infants. Early humans colonized a new and very risky evolutionary niche. They took a bet on large brains (or rather natural selection took that bet for them) and paid a significant behavioral price for doing so: the need to nurture their offspring for an exceptionally long period.

What were the large brains for? In a word, cooperation, of a scale and complexity that no primate had yet attempted. The more sophisticated the relations of cooperation and reciprocity within a group of individuals, and the larger the group of individuals concerned, the greater the cognitive challenge of keeping track of mutual obligations. Among primates, species with larger brains (relative to their body size) tend to live in larger groups.[29] This suggests that larger brains are a necessary cost of living in larger groups (one that implies an increasing selective pressure in favor of larger brains as group size increases). It also suggests that the benefits that accrue to primates with larger brains include those that can be realized only by exploiting the cooperation possibilities in larger groups. Today human societies are based on elaborate networks of cooperation that span the entire globe, but our hunter-gatherer ancestors had no vision of such a consequence, and of course natural selection could have had no foresight of it either. Cooperation even within hunter-gatherer bands expanded the range of natural environments that could be exploited and foods that could be consumed, and these benefits of cooperation were large enough to offset the very important costs.

Large brains are expensive: they need a lot of protein to build and a lot of energy to run. Human young (and their pregnant or nursing mothers) need more meat and more calories than do the offspring of young chimps to supply their growing brains. Foraging for such a diet requires more ambitious social arrangements: greater cooperation among hunters and a greater willingness for gatherers to move large distances when necessary in pursuit of food. Careful studies of forager communities by modern social anthropologists have shown that their young members typically do not produce enough food to feed themselves until they are in their late teens. Although adolescents are strong and nimble, both hunting and gathering need skill and experience, and adolescents are usually big eaters. The time of life when foragers are producing the biggest food surplus is not in their early twenties, when they are strongest, but in their early forties.[30] Children and adolescents depend not just on prime-age adults but also on older people, especially grandmothers, who contribute surprisingly large amounts to their overall nutritional needs.[31]

In addition, as Richard Wrangham and his coauthors have argued, the development of cooking brought about a radical shift in human social arrangements. Cooking enabled the extraction of much higher nutritional value from the foodstuffs human beings could find, but the time it required made these foodstuffs more vulnerable to theft. Human males were co-opted as protectors by females in exchange for the superior food—and also for the promise of a greater stake in the family unit. So eating cooked food both required and permitted more cooperative social arrangements than any primate society had previously seen.[32]

The second cost is that large brains require large skulls to house them. At the same time that natural selection was favoring infants with larger skulls, it was also selecting among women for those who could stand upright and move quickly on two legs. Given the basic model of the human body on which natural selection had to work, bipedalism set a limit to the size of the pelvis: a woman with a narrow torso and wide legs would have waddled much too slowly away from her predators. Although hominin females grew in size over the course of our evolution (more than males did, in fact), the metabolic costs of a larger overall body size prevented them from growing proportionally to their babies' skulls. The skull of a baby at the end of pregnancy can only just fit through the space in the pelvis. The only way for natural selection to make human babies with larger skulls was to end gestation prematurely and allow them to be born in a state of dependence that no other animal (except marsupials) could possibly manage. So to the costs of feeding large-brained babies must be added the costs of protecting them, and protecting them requires sophisticated social arrangements. Our human ancestors took a very large gamble on those large-brained babies, a gamble that depended on their being able to realize cooperative benefits that would outweigh the costs. It was a gamble that nearly failed. Most branches of the hominin line died out, including the Neanderthals, who had slightly larger brains than ours. Only our branch survived into the modern era, and it too very nearly didn't make it. Humans needed to adapt to new habitats many times, until eventually those large brains gave them the flexibility and inventiveness to change the habitats themselves.

This direction of evolution might appear to have been good for women, because men were now contributing resources to their offspring—notably food and protection—that they had not contributed when our ancestors were more like bonobos and chimpanzees. In some absolute sense it probably was good for women. But it certainly changed women's bargaining power relative to that of men, and changed it for the worse. For if men were contributing more, their contributions also became more essential, both because the contributions were more necessary (protein was important for those growing infant brains), and because fewer men among those available might be counted on to provide it, given the longer-term nature of the commitment involved. Being essential enabled men to exact a price for their contributions. In a word, the resources controlled by men had become more scarce.

Scarcity and the Gender Bargain

An unequal exchange doesn't have to be the result of coercion: it can be the outcome of bargaining between individuals who have different degrees of control over the resources at stake. In most hunter-gatherer societies, men hunt for meat, and women gather roots, fruit, berries and so forth. There has been controversy over whether men and women contribute equally to subsistence in such societies, but in some sense that controversy is irrelevant, because there's no particular reason to expect the terms of the exchange to have been very favorable to women. Human beings need high-protein diets (especially for infants with growing brains). Men may or may not be effective at gathering calories (anthropologists disagree on this point),[33] but they're the specialists at hunting large quantities of protein. Women can't usually hunt, not because of any lack of ability but because hunting would be too dangerous for their infants, or simply impractical in the circumstances. As a result, the men have a resource the women need, but the reverse is not true: men can get calories on their own if they have to. The men's resources are therefore more scarce and consequently command a higher price. There's nothing moral about this: it's simply the outcome of the bargain.

Consider a simple example. The point of the example isn't to describe literally how a hunter-gatherer society would have operated, any more than the point of a map is to look like the landscape it helps you to navigate. It's rather to see how men could have gained an advantage just by the fact that they could produce something that women needed, even if they hadn't also had other advantages, such as the power to coerce women physically. Suppose in a group there are ten men and ten women. Each of the ten men can gather one kilogram of starch (tubers, say) in a day. Let's suppose the women are more productive than the men: each one can gather two kilograms of starch in a day. However, the group of ten men can also hunt, and once every ten days or so, they bring back an animal that yields one hundred kilograms of meat, equivalent to one kilogram of meat per man per day. The men can choose whether to gather starch or to hunt meat. As it happens, both meat and starch are essential for life: each person has to eat, on average, a minimum of a quarter kilogram of meat and a half kilogram of starch every day. But let's suppose that meat is strongly preferred: if meat and starch cost the same, everyone would prefer to eat three times as much meat as starch. Let's assume that their relative preferences for meat and starch don't change as their overall intake increases (that won't be strictly realistic, since preferences for meat are more characteristic of better-fed people, but it makes the example simpler to understand).

The men decide to see how much starch they can get in exchange for their meat if they bargain with the women. They decide to negotiate a rate at which each person can exchange what they have against what they want. Before the bargaining, each man has one kilogram of meat, and each woman has two kilograms of starch. You might think the obvious rate to exchange would be one for two: half a kilogram of meat for every kilogram of starch, so that each person would give up half of what they started with, and everyone would end up eating half a kilogram of meat and one kilogram of starch. But remembering how much people prefer meat, at that rate of exchange, everyone would rather give up half their remaining starch in order to get half as much meat again. So everyone will want to get more meat for less starch, until it occurs to someone to offer a better rate, say three kilograms of starch for every kilogram of meat.

In fact, under some reasonable assumptions about how the need for meat increases as you have less and less of it, the rate at which meat exchanges for starch might rapidly go up as high as six to one! Given that everyone prefers meat enough to want to eat three times as much of it if the prices were equal, it takes a price ratio of six to one to induce them to eat only half as much. At that rate, after the bargaining the men will end up with three-quarters of a kilogram of meat and one and a half kilograms of starch, while the women will end up with one-quarter of a kilogram of meat and just half a kilogram of starch. Only at these values will the high price of meat finally restrain their desire to eat by enough to make them ration out the small amount of meat available. But the result is startling. The men have three times as much as the women, even though the women are twice as productive as the men at the one task at which their performance can be compared, which is the gathering of starch. And the men only really work one day in ten.

So why does it work out that way? It's because, in this scenario, meat is essential to life, it is highly desired, and men have all of it. Of course, starch is also essential to life, but it is less highly desired, and women don't have a monopoly on the starch (men can get it too, if they want). So meat is *scarce* in a way that starch is not, and for this scarcity men can extract a price. No physical compulsion is needed, just a hard bargain. If instead people could just substitute in their diet a kilogram of starch for every kilogram of meat, then meat and starch would exchange at an equal rate—nobody would pay more. Then women (who start with two kilograms) would end up eating twice as much as men. It's the fact that they don't have access to a substitute for the men's meat that puts them at a disadvantage, not greater merit on the part of the men.

Of course, real life is (and always has been) different from this example in many ways. Some of these differences make the outcomes less unequal: for instance, men and women are not always driving purely selfish bargains with each other but often care about each other's welfare. Nor are they always bargaining for food for their own consumption: they typically obtain food to share with children and other family members, and often with group members outside their immediate family as well.[34] Another major difference is that, as Sarah Blaffer Hrdy has repeatedly emphasized, in most societies mothers receive help in looking after their infants from a whole network of adults other than the father (or presumed father). Siblings, grandparents, and unrelated adults can all play a part, and together they reduce the extent to which mothers are dependent on any particular man and even on the men in the group as a whole—which is particularly fortunate given the notorious unreliability of that dependence: "Does this mean that fathers are not important? No. However, it does mean that a mother giving birth to slow-maturing, costly young does so without being able to count on help from the father. The impact on child well-being of variable paternal commitment depends on local conditions and on who else is around, able, and willing to help."[35]

However, there are also some ways in which real life differs from this example that tend to reinforce rather than diminish the inequality between the sexes. Most important is the fact that men are not just negotiating individually but often collude. In particular, they have colluded in the coercion of women and have often ostracized or physically assaulted those women who try to escape the coercion. This collusion has been a natural by-product of the increasingly sophisticated collaboration undertaken by males in all societies for a range of purposes, including food production and warfare. (Gathering can be cooperative but doesn't have to be, or not to the same extent.) Those activities have reduced the autonomy of females, as we can see even among the other great apes. For instance, female bonobos can assert themselves more successfully against males than can female chimpanzees. This difference appears to be due to differences in the distribution of food in their foraging range, as a result of which male bonobos have fewer reasons to cooperate than male chimpanzees do. Hunting in groups, for instance, is more common among male chimpanzees (females are at a disadvantage in hunting because of their infants).

Cooperation in one domain, such as hunting, can extend easily to another domain, such as cooperating in the sexual harassment of females or in their confinement under the watchful eye of their mates. Females face a disadvantage in cooperation because, in all great ape species, the adolescent females leave the group to mate and so are unlikely to find themselves in a group with other siblings (there's some evidence that this was true of most prehistoric human groups as well, though the ethnographic evidence for surviving groups that have been studied indicates a considerable amount of flexibility in residence patterns).[36] This is a disadvantage that bonobos have managed to overcome in part by their exuberant use of sexual play both to reinforce ties between females and to defuse the aggression of males. In *Homo sapiens* (a less sexually inventive species), though it's true that harassment and confinement of women were less common among hunter-gatherers than they later became under agriculture, that would certainly not have been for want of trying on the part of the men.

There are many factors other than bargaining, then, that have affected the balance of power between men and women. Nevertheless, our example shows that to the extent that they do bargain, the result can be strikingly asymmetric because of the luck that gives one party the control of a particularly scarce resource that the other needs. An important reason why human males have exerted so much more control over human females than chimp or bonobo males exert over females of their own species is simply that human males have had things to offer that human females needed much more than the females of those other species do, namely food and protection. And they needed these because natural selection took *Homo sapiens* into an evolutionary niche dependent on big brains and the capacity for social cooperation. That looks like bad news for women everywhere, but it's actually good news for modern women. For, as I show later, the balance of scarcity, with all that it implies for dependence, can be changed—up to a point.

Goodbye to the Hunter-Gatherer Life

The ecological conditions under which most hunter-gatherers lived imposed a fairly strict constraint on the extent to which men were able to harass and confine women. Women needed a fair amount of autonomy in order to forage. Shares of food consumed were also probably much less unequal than our example suggests they could have been under purely selfish bargaining. There is evidence, for instance, that in many societies, males have competed to be seen as good providers, even if they have sometimes chosen ways of demonstrating their prowess that privilege display over nutritional efficiency (hunting large game, for instance, may be a less efficient way of finding protein than going after small game).[37] The inequality that mattered, though, for the degree of autonomy enjoyed by women, was not in how much men and women consumed, but in how much each of them controlled.

Relations between men in forager societies were almost certainly fairly egalitarian, because hunting requires trust rather than compulsion, and attempts to coerce others mean that they will simply leave. Christopher Boehm's book *Hierarchy in the Forest* has argued that this was not because *Homo sapiens* lost the competitiveness and the status consciousness that characterizes ape societies but because individuals who abused their power and status would provoke a counterreaction by coalitions of the weaker members of their group.[38] In a hunter-gatherer society, the cooperation of these weaker members was essential even to the strong. Whatever the roots of this comparative egalitarianism among men, the implications for women were profound. They enjoyed some freedom to choose their men, and the choice was not (as it would become later, if they had any choice in the matter at all) between becoming the only wife of a very poor man and the junior wife of a very rich one.

It's important all the same not to romanticize the hunter-gatherer life. It was highly violent by today's standards: the best current estimates place the proportion of deaths from violence at around 14 percent, roughly ten times the rate for the world as a whole in the twenty-first century.[39] It was also unhealthy: skeletons of hunter-gatherers from North and South America reveal that more than half of them suffered from abscesses that must have caused agony, and the state of their teeth is painful even to imagine.[40] On both these counts, though, farming was to make things at first a great deal worse.[41] And the conditions of women were to deteriorate by even more than those of men.[42] Men were to use their talent for cooperation—in particular, their talent for devising and enforcing complex systems of social rules—to restrain and confine women much more than had been possible during the long ages of hunter-gatherer life.

It is one of strangest paradoxes in the natural world that females, endowed with scarce biological resources (their eggs), should have become so powerless in the face of the males who started out as impoverished, controlling only their cheap and abundant sperm. If, as I've suggested, it was men's good luck in controlling the scarce meat needed by women that enhanced their bargaining power, how can it also be true that women have been weakened, not strengthened, by possession of the scarce gametes? The answer lies in the difference between assets that you control securely and assets that control you. It's the difference between having a million dollars kept in a safe place and having a gold filling worth a million dollars in your teeth. The first you can use at your leisure and free of any fear of coercion. But to turn the second into anything valuable you have to find a cooperative dentist, and in the meantime you need to be careful about where you go walking at night. Men's resources, when they have had them, have been ones they controlled; women's resources have often controlled them, because they were physically vested in women's bodies. In short, the value of scarce resources depends not just on how scarce they are and on how much people want them, but also on the ease with which they can escape them and the degree of security they enjoy in their use of them. And that in turn depends on the framework of rights and conventions that govern their use. For our hunter-gatherer ancestors, those rights were entirely informal and largely consensual. Agricultural societies were to redefine them in ways that were not usually to women's advantage.

It is also a paradox that the developments that led men to restrict women's autonomy so severely grew out of the most remarkable experiment in social cooperation that the world had ever seen. The adoption of agriculture transformed not just the technology of food production but humans' whole way of life. It obliged them to settle so as to guard their fields; it brought them together into villages, towns, and cities; it obliged them to defend themselves; and it created a surplus that made it possible for farmers to pay for the activities of soldiers, scribes, priests, and kings. Jared Diamond has called it the greatest catastrophe in the history of humanity; it was also, without doubt, the foundation of all modern civilization.[43] Charles Darwin was in no doubt that human social cooperation had emerged because of, not in spite of, natural selection. And for him, sexual reproduction was central to that cooperation, indeed emblematic of it. To the man who famously set out a balance sheet of the advantages and disadvantages of marrying, and who depended for his researches on a network of collaborators who supplied him with observations of natural and social phenomena from across the globe, the study of sexual reproduction was a moving reminder that nothing worthwhile in life can be accomplished entirely alone.

PART TWO

TODAY

From prehistory we jump forward to the present day, and that shift warrants an explanation. This book is about the traces our evolutionary past have left on the economic relations between men and women in the twenty-first century. The brains and bodies we have inherited from our early ancestors are navigating a world very different from the one in which they first evolved. This book is not about how our evolutionary past affected life during earlier periods of history. Nor is it about how those earlier periods of history have left their traces on us today, except in one respect. The tug of the historical past may explain some of the reasons why men and women sometimes continue to play different roles in important parts of modern economic life. But when I conjecture that the tug of the past is the explanation, I am no more specific than that; and my evidence is usually that other explanations will not do. Writing in detail about that historical past would be a fascinating exercise, but it is not the task of this book.

Drawing conclusions about prehistory is very different from doing so about modern society, and the chapters in part two have a different feel from those that have preceded them. About prehistory we have very little information: recovering a picture of that past is like assembling a jigsaw, most of whose pieces are missing forever. About the present we have an avalanche of information, most of it irrelevant at best and misleading at worst. Creating a picture of the present is like assembling a thousand-piece jigsaw from a million possible pieces; the trick is learning what to leave out. Statistics are a vital tool in helping social scientists to decide which facts to ignore, just as maps are a vital tool in helping travelers decide which parts of the landscape are irrelevant. Statistics can tell you, for instance, that there's no point attaching too much significance to a particularly charming anecdote because it really isn't representative of the overall picture. So whereas the argument in chapters 1–4 is defiantly, even joyously, anecdotal, the argument here is no less determinedly (and also no less joyously) statistical. I deal with such methods as psychometric testing and regression analysis, though nothing I write presumes any prior technical knowledge about these things. As it happens, psychometric testing and regression analysis are alternative ways of doing the same thing: looking at a complex, multidimensional reality and trying to reduce it to a single number that captures what you care about for the purpose at hand. Different purposes require different numbers, so no one procedure is ever the right one independently of what we want to use it for. You can ask various questions about a book (how many pages it has, what it weighs, what it costs, and how often you have yawned while reading it, for instance). While any of the answers to these questions may matter for your current purposes, they all leave out information that in some other contexts you rightly regard as important (all of these pieces of information about the book, for instance, omit any mention of anything the book actually says).

The first half of this book describes how human males in prehistory developed ways of accumulating economic resources in order to advance their interests in a sexual bargain. The circumstances that allowed them to do this are no more. Indeed, in modern industrial societies, the entire basis for a sexual division of labor in principle has disappeared, outside a tiny number of special occupations (mainly military ones), and even there, its continued presence remains controversial. In practice, that division of labor has greatly diminished, but it has not disappeared, and the occupations in which women are least represented include many of the most powerful occupations in modern society. The question that occupies the second half of the book is therefore whether the inequalities of economic power that characterized our hunter-gatherer ancestors persist to some degree in the present day; and, if so, why.

The question is not whether men are healthier or happier or more fulfilled than women, nor even whether they get to consume more economic resources than women do. The male dance flies we saw in chapter 1 acquire food packages in order to control females, and even if it is the females that end up consuming the food, it is the males that thereby exercise control. There's no evidence that women in the modern world consume less than men relative to their needs or are less healthy as a result. Indeed, such evidence as we can find tends to point in the opposite direction: in the world as a whole, men suffer 20 percent more deaths from infectious disease and nearly twice as many deaths from injuries and violence (deaths from the other main category, noncommunicable disease, occur in very similar numbers among men and women).[1] The only countries in the world for which the World Health Organization reported lower life expectancy at birth for women than for men in 2009 were Tonga and Tuvalu.[2] Nor are these comparisons a feature just of the modern world. In forager societies, as far as we can tell, mortality among girls under five has often been higher than mortality among boys, but provided they survive the first five years or so of life, women have lived longer and somewhat healthier lives than men.[3]

Evidence on happiness is harder to interpret, not least because of widely divergent views on what constitutes fulfillment for men and women, but I do not want to claim that women are systematically more unhappy or less fulfilled than men, though other researchers have indeed made this claim.[4] It is certainly true that in many countries, women are subjected both to illegal but inadequately policed violence and to legal violence for reasons related to their sexuality (though they typically suffer less violence than men from other causes). By legal violence I mean laws that impose on women all kinds of physical trauma, from genital mutilation performed in an attempt to stifle their sexuality before it has a chance to express itself to capital punishment for alleged offenses against sexual modesty. Although there are things to be said about how our prehistoric heritage has brought these terrible practices into being, they are less pressing than the question of how such violence can be stopped; but this topic would require a separate book. Here I am interested in a quite precise and quite different question, which is whether the masculine control of economic resources that has characterized human societies since prehistory is about to disappear.

The fact that some men have exercised control over a large share of the world's economic resources does not imply, and has never implied, that all men do so. Indeed, at the same time that men at the top of society have enjoyed privileges beyond any to which women could aspire, there have always been men at the bottom who have done much worse than most or all women. To take the most enduring criterion of success (indeed, the only one to which natural selection pays any attention), the fact that some men have been able to have many more children than even the most fertile women implies (inevitably, because every child has one mother and one father) that at the bottom of the fertility distribution, many more men than women have died leaving no children at all. This also implies, incidentally, that the pool of men from whom the current human species is descended is smaller than the pool of women, a fact corroborated by studies comparing DNA of current human populations in the Y chromosome (transmitted only by men) and the mitochondria (transmitted only by women).[5] In all known human societies, more men than women have been homeless or in prison (these conditions being often both a symptom and a reinforcement of their poor fertility prospects).

Recently this long-standing and well-understood fact about human societies has been rediscovered and rebranded as the "crisis of men" or the "decline of men."[6] It has been pointed out, correctly, that far more men than women are in prison, that far more men than women are homeless, that more men than women are unemployed, and that there are now substantially more women than men enrolled in higher education in the United States (as well as an increasing gap between men's educational attainment and the higher levels attained by women).[7] Of these four statements, the first two have been true for as long as anyone has been able to measure the prison and homeless populations, and it's not clear that changes in numbers of imprisoned and homeless men over time tell us anything about men in general as opposed to the incarceration policies of various governments in particular. The third (the unemployment gap) has been a feature of the current recession in the United States and the United Kingdom, but after an initial leap the gap between men's and women's unemployment is declining in both countries, and it is far too early to tell whether it will persist in the longer term. In any case, the unemployment gap is not a feature of continental Europe, and there is no evidence that it portends the permanent redundancy of men.[8] The fourth statement (the education gap) is true because women's college enrollment rates have been rising, not because those of men have been falling, so it represents evidence for a crisis of men only if you think that education for women must be bad news for men—a point of view that can be defended but is far from being a foregone conclusion.[9] I return to the crisis of men in chapter 8, but for the moment I just want to note that none of these facts shows that the control by men of society's economic resources is an unreal or an unimportant phenomenon. And it is this phenomenon that is the subject of part two.

I ask whether, in the modern world, women can now aspire to control economic resources in proportion to their talents. Not everybody would agree that this is a reasonable standard of evaluation: some people, for instance, believe that talent should not be a basis for entitlement to economic rewards. But it's the question I deal with in the first three chapters of part two, and so it is important to understand what I shall mean when I talk about talent.

What then do I mean by talent? Talent is not some mystical spirit in people, a kind of genius lite, but just a name for an assorted jumble of different capacities for behaving in reliably systematic ways—abilities, in short. People have many different abilities, and the ones that count in what follows are the abilities they have for doing things that other people value.

Human life is founded on exchange, and all of us, barring some truly exceptional cases, live by doing things for others, which induces others in turn to do things for us—things without which our lives would be nastier, even more brutish, and considerably shorter. Some people are lucky to have abilities that others value a lot, while others are unlucky and get substantially less in exchange for their own effort. I define *talent* as just the fact of having abilities that, overall, other people value. (Talent in this sense is not necessarily a virtue, nor even something we should always admire: it's just a fortunate fact.) Sometimes two individuals earn different rewards because they have different talents: one is lucky enough to have the ability to make or do things that other people value very highly. Sometimes it is because, although the individuals have very similar talents, the process of economic exchange causes one to receive a higher share of the value that their activity creates for other people. We saw in chapter 4 that this share can vary for a number of reasons: for example, it can vary because one individual faces more competition than the other (one may be the only plumber in a small town while the other is one of many in a large town).

In the next three chapters I show that other factors may influence the share as well: one person may be good at negotiating or at signaling her talent while the other, no less talented, may negotiate less effectively or operate in an environment where talent is not so easy to signal credibly. I ask whether the differences in economic rewards between men and women that persist in the twenty-first century do so because they reflect different talents or because men and women tend to receive different shares of the value that their talents create. Women and men have sexual favors they can trade, and in prehistory that trade also determined the terms on which they traded everything else. The question now is whether that exchange of sexual favors continues to determine the terms on which their other talents are traded, or whether the two sets of exchanges have at last become uncoupled. If they have, that would be a development without precedent in the history of our species.

To develop my argument in terms of talent, I start in chapter 5 by talking about something other than gender altogether. I return to gender very shortly, but first I want to consider a different way in which we might have chosen to divide the human species into two different castes.

FIVE

Testing for Talent

No man is sure he does not need to climb.
It is not human to feel safely placed.
"A girl can't go on laughing all the time."
—William Empson, "Reflection from Anita Loos," 1940

A Tall Story

IMAGINE THAT YOU LIVE in a society where most of the important decisions are taken and the most important rewards enjoyed by people who are tall. Actually, you don't have to imagine very hard, because there's a good deal of evidence that modern industrialized societies do grant substantial economic power and privileges to tall people (more on that evidence below). But suppose height were even more central to economic and social life than it now is. Suppose, for instance, that every time you filled in a bureaucratic form you had to declare your height, and there were separate toilets and changing rooms for shorter and taller people. Suppose that endless daily rituals kept underlining the distinction for you: hosts deciding seating plans at dinner always insisted on alternating short people and tall people; clothes shops devoted separate sections to short people and to tall people, sporting significantly different colors and styles; and special magazines appeared with titles like *Tall Tales* and *Short and Sweet*. And suppose you discovered that there were large discrepancies in the representation of short and tall people in positions of power and influence: that, say, among the chief executives of major companies, only one person in thirty was of below-average height. What evidence might it take to persuade you that this was a reasonable, fair, and economically efficient state of affairs?

One possibility is that someone might offer you evidence of ability, based on various psychometric tests. Suppose it were clear that, on the basis of the tests currently available, tall people performed on average better than short people from very early in their schooling. The apparent discrimination against short people in employment and earnings might just reflect the fact that short people were, on average, less talented—at least as measured by the tests. Would that be enough to justify the outcome? Clearly not. It would depend on at least two further factors: whether the tests were actually identifying abilities relevant to the people's later performance in employment, and whether the degree of discrepancy in test performance was large enough to explain the later discrepancy in economic rewards.

Imagine, then, that as you inquired into these tests that children take in school, you discovered that in addition to some tests of arithmetic and verbal comprehension, there were also tests of athletic ability, such as performance on the high jump and scores in basketball. What was presented as an "ability score" was just some aggregate of performance on a group of these different tests, some of which just seem to be ways of favoring tall people from the start. Don't worry, you might be told reassuringly, statisticians have established that these tests are justified because they have a very high power in predicting individuals' subsequent success. At this point the shade of Franz Kafka might begin to show some restlessness: the economic outcomes are justified because they are predicted by the tests, and now you're told that the tests are justified because they predict the economic outcomes? This clearly won't do.

In fact, it wouldn't do even if you could be convinced that the tests were both positive indicators of real underlying talent and positive predictors of economic outcomes. Two factors (in this case, talent and economic outcomes) are not necessarily related just because both of them happen to be correlated with some third factor (test scores). If you're doubtful about this, consider a similar faulty inference: how fast you drive may increase your chances of arriving early at your destination and also increase your chances of having an accident, but having an accident does not increase your chances of arriving early at your destination. In other words, correlation is not transitive: factor A (arriving early) can be positively correlated with B (driving fast), and B can be positively correlated with C (having an accident), without A's being positively correlated with C. That is because the aspects of B that are correlated with A are not the same as those of its aspects that are correlated with C.

To avoid this circular reasoning, someone who wanted to use the predictive power of psychometric tests to defend the economic rewards enjoyed by tall people as being a just return on talent would need to argue that the features of psychometric tests that enable them to reflect talent are exactly those features that help to predict economic outcomes. Because talent and economic outcomes are both complex, multidimensional variables, that is not a straightforward exercise. It would require ruling out two possible causes of the correlation between test scores and economic outcomes.

We could call the first of those causes *irrelevant inclusion:* test scores might be influenced partly by talent but also partly by some other unrelated characteristics, which are themselves unjustifiably rewarded in economic outcomes (such as the ability to play basketball or to do the high jump). There could indeed be a correlation between test scores and economic outcomes, but one that had nothing to do with talent. We might call the second possibility *unjustified exclusion:* test scores reflect only some aspects of talent, and the aspects they fail to measure might also be unjustifiably ignored by the process that determines economic rewards. The correlation between test scores and economic outcomes could show that both were reflecting an incomplete and inadequate measure of talent. In both cases the argument would need to be based on detailed examination of the tests, not just on the correlation of the test scores with the outcomes.

As it happens, researchers have done exactly that. As I hinted at the outset, the idea that tall people gain a more than equal share of economic power and rewards is not a fiction. It's been known for around a century that tall people earn more and are more represented in high-status occupations. The social psychologist E. B. Gowin wrote in 1915 that business executives tended to be taller than "average men"; he also compared the height of people with different status in the same profession, noting that bishops tended to be taller than small-town preachers, sales managers were taller than salesmen, and so on.[1] The economists Anne Case and Christina Paxson have shown that the correlation between height and economic success remains very strong today: for instance, they report that "an increase in U.S. men's heights from the 25th to the 75th percentile of the height distribution—an increase of 4 inches—is associated with an increase in earnings of 9.2 percent"; and they report similar findings for women.[2] The explanation they propose is that "the height premium in earnings is largely due to the positive association between height and cognitive ability, and it is cognitive ability rather than height that is rewarded in the labor market."[3]

To prove their point, as Case and Paxson know, it's not enough to demonstrate that test scores are correlated with height and height with labor-market outcomes. Though they don't express it this way, they also need to ensure that the correlation with test scores is not due to either irrelevant inclusion or unjustified exclusion. For instance, they show that height in children at ages 5 and 10 is strongly correlated with scores on a range of tests of different cognitive skills, ranging from figure drawing to linguistic skills and mathematical ability.[4] The probable reason is that height and cognitive skills are both the outcome of factors enhancing the growth and development of young children.[5] This makes irrelevant inclusion unlikely: if the correlation of scores with height were due to some extraneous factor unrelated to cognitive ability (like jumping or basketball in our hypothetical example), it's unlikely the extraneous factor would be present to the same degree in all of these different tests. Unjustified exclusion is harder to rule out for certain: maybe some important talents that are more strongly present in short people are being overlooked by these tests. But in the absence of plausible theories about what these talents might be, the wider the range of tests that are used, the less likely it is that such talents are being overlooked. All in all, therefore, it does seem as though the correlation between test scores and height does reflect a genuine causal relationship between talent and height.[6] But is that causal relationship important enough to account for the fact that tall people receive, on average, so much more generous economic rewards?

The answer is no. Interestingly, Case and Paxson don't find that the effect of height on salaries is due entirely to the causal relationship between height and talent. Using data from the United States, they find that controlling for

Short and tall men. Tall men earn significantly more than short ones. © Anna Peisl/Corbis.

test scores reduces the return on height for men in the labor market but does not eliminate it completely.[7] The effect of height on earnings falls to about half of its previously estimated level when test scores are taken into account, meaning that fully half of the effect of height on salaries remains unexplained by cognitive ability. In the data for women, the measured return on height is reduced: it remains positive but no longer statistically significant (in their data from the United Kingdom, this is true for both women and men). It's clear that a substantial part of the measured return to height in the labor market is really due to underlying cognitive ability. But it seems likely that height also continues to convey some additional advantage. Whether that's because of a prejudice in favor of height or because height correlates with other characteristics that employers value besides cognitive ability, or because (perhaps unconsciously) employers use height as a more reliable signal of talent than it actually is, this evidence cannot tell us.[8] And in case you were wondering, the difference in earnings between women and men has nothing to do with the difference in their heights. Whether you control for height or not, women face the same salary disadvantage compared to equivalently talented men.

Gender, Talent, and Rewards

Let's return, then, to the issue of gender, which poses a similar puzzle about the difference in economic rewards. The period since 1900, and especially since the Second World War, has seen industrialized countries, as well as many developing countries, remove formal barriers to women's participation in almost all areas of employment. These countries have also revoked or outlawed many barriers that were based on explicit custom and practice. This is a remarkable experiment, representing the first large-scale attempt in the history of the human species to remove obstacles to the division of labor between men and women and to allow members of both sexes to perform (almost) any kind of work they can persuade someone else to pay them to do. During prehistory, men and women appear to have occupied largely different roles (though, because it was prehistory, we have no written evidence and have to make inferences on the basis of other kinds of evidence, as we saw in chapter 4). Throughout much of recorded history they occupied largely different roles because barriers to their doing otherwise were enforced by those groups that held economic and political power, which consisted overwhelmingly, if not quite universally, of men. Most merchant guilds in medieval Europe, for instance, refused entry to women or permitted it only to the widows of deceased members and then only under fairly restrictive conditions.[9] The landscape was not uniform, and few of the barriers were absolute: examples of women entering apparently male occupations can be found in many countries, but the formal and informal obstacles they faced were often daunting.[10] Many of these barriers persisted long after mechanization had neutralized any advantage that men had once possessed in physical strength.

Since the removal of such explicit barriers, women have begun to occupy many of the roles formerly occupied only by men. This period coincided with technological developments such as the invention of household labor-saving devices[11] and the contraceptive pill, which reduced significantly the costs to women of participating fully in the labor market and pursuing the studies necessary to make that participation possible.[12] It's also not a coincidence that this change came soon after women gained the vote, which in most of the industrialized countries occurred after the First and Second World Wars, the first conflicts in which women's contribution in previously male occupations such as munitions manufacturing had proved essential to the war effort.[13] (Famously neutral Switzerland was the last republic in the Western world to grant women the vote in national elections, in 1971.) Changes in technology, changes in the law,[14] and changes in attitudes all reinforced one another, with changes in attitudes given particular visibility by such signals as women wearing trousers and smoking cigarettes (dubbed "torches of freedom" by the American Tobacco Company in 1929).[15] Women could not only manufacture chemical weapons every bit as well as men; they could inhale them, too.

Whatever the exact configuration of causes, by the standards of previous historical periods, the resulting pace of change has been spectacular. In 2008, just under 60 percent of US women of working age were in the labor force. Although still lower than the 73 percent participation rate of men, this figure was a dramatic change from 43 percent in 1970 and just over 30 percent in the late 1940s.[16] Of women with children above the age of five, 77 percent were in the labor force by 2008.[17] Not only have women entered the workforce in historically unprecedented numbers, but they have entered hitherto male-dominated professions, from accountancy to zoology, and in many of them have come to equal or outnumber men in the space of a few decades. In 2009, women represented 51.4 percent of what the US Bureau of Labor Statistics calls "management, professional and related occupations." In the United States women also significantly outnumber men in enrollment in higher education, which is the main port of entry into interesting and remunerative work for most of the population. In short, an unprecedented tide of talented and motivated women swept into the masculine economy, producing an incalculable, unrepeatable, but surely vast increase in activity, output, and innovation in many areas of social and economic life.

And yet that tide did not sweep everywhere, and it did not lift all boats. The very speed of this dramatic change throws into sharper relief those areas where change has been slow or nonexistent. Three things remain highly puzzling. First, some occupations continue to see low proportions of women even though their formal barriers are apparently no higher than in occupations where women have attained parity. Women represent only 32 percent of lawyers, 25 percent of architects, 20 percent of computer programmers, 15 percent of taxi drivers and chauffeurs, 7 percent of civil engineers, 2.2 percent of electricians, and 1.3 percent of airline pilots. Only 32 percent of physicians and surgeons are women, even though women make up nearly three-quarters of all health care practitioners.[18] There are some differences between the United States and other rich countries (notably because overall female participation in the workforce is lower in countries such as Italy, Japan, and Germany), but broadly similar patterns are observed across the industrialized world.

Second, women's salaries continue to be lower than men's even within occupations. The overall ratio of US women's earnings to those of men was 81 percent in 2010, and in some professions it was much lower (77 percent among lawyers, for example, and 58 percent among personal financial advisers).[19] Third (and related), across a broad range of economic activities, many of the most prestigious and highly remunerated positions continue to have startlingly low rates of representation of women. In 2010, women made up only 15.7 percent of board members and just 2.4 percent of chief executive officers of Fortune 500 corporations.[20] There's some evidence that when women are appointed to leadership positions, these tend to be more precarious ones (this phenomenon has been christened the "glass cliff").[21] Even within a specific area of activity, such as the restaurant and catering industry, women make up more than 70 percent of waiters and waitresses and 41.5 percent of cooks but only 20.7 percent of chefs and head cooks.[22] What is going on?

The story of height might make you think I'm going to suggest that the explanation might be differences in talent. But the point of the analogy with height was to show that using measures of talent to explain differences in rewards requires that these measures satisfy a number of rather particular conditions. Outside a small number of occupations for which physical strength remains important, there's simply no evidence at all that the underrepresentation of women in certain occupations, their lower salaries, or their underrepresentation in particularly high-status or well-rewarded positions has anything at all to do with differences in talent between women and men.

We can start by looking at psychometric tests of cognitive ability, or skill. These do not test a single type of skill but many, usually through the administration of a number of component tests that form part of an overall package or battery. There are some differences between men and women, on average, in their performance on these different tests of skill: women tend to perform better on average than men in tests of verbal ability, and men tend to perform better on average than women in tests of some (though not all) types of visual and spatial skills.[23] These average differences are, however, usually small compared to the variation between individuals of either gender, and some of them are found in certain environments but not in others.[24] There has been vocal and justified criticism of the specious precision of some of the tests, which are known to be influenced by contextual factors such as the information given to subjects before the tests are administered.[25] In particular, the relative performance of women has been shown to be subject to what is known as "stereotype threat": when primed to be aware of stereotypes of poor female performance, women perform less well than when primed with neutral messages.[26] These considerations mean that a lot of uncertainty surrounds the evidence on gender differences in performance on these component tests, but for the moment it seems reasonably likely that at least some such differences are real.

Psychologists speak of *g*, a measure of general intelligence, as a persistent common component of all reputable psychometric tests; it is also highly heritable and highly correlated with economic performance.[27] This heritability does not mean, incidentally, that environmental influences on *g* are unimportant. On the contrary, average IQ scores have risen rapidly over time in most countries around the world, by around three IQ points per decade. This increase is known as the Flynn effect, after the researcher who first documented it, and it is entirely implausible that it could be due to genetic change.[28] It therefore indicates that the skills captured by IQ tests are strongly influenced by environmental factors that have been changing over time, though to date there is no hard evidence as to what these might be.[29]

There is a large and contentious literature debating whether *g* shows any significant average difference between women and men. Some studies find no difference, some find a difference in favor of men (equivalent to between one and five points on a standard IQ scale), and a much smaller number find a difference in favor of women.[30] The male advantage, where it is found, is equivalent to only about a decade's worth of the Flynn effect and could therefore be due entirely to differences in the learning environments for boys and girls, if these are as great as differences in the learning environment from one decade to the next. Within this margin of variation, such findings are simply irrelevant to the question of whether differences in talent sufficiently explain men's and women's differential performance in the labor market. We would not learn anything from knowing that the average difference was "really" zero, or 3 percent in favor of men, or 1 percent in favor of women.

The reason is very simple. Precisely because there are differences on average in men's and women's performance on the component tests, the common factor of talent that such tests appear, statistically, to reveal, will be sensitive to the choice of component tests that make up the overall battery, and in particular whether there are more of them that tend to favor women's performance or more that favor men's. In other words, the fact that all batteries of tests can be used to derive a measure of *g* should not make us think that the measures of *g* they produce are identical. On the contrary, specific estimates of *g* that particular tests produce can rank a group of individuals very differently according to the types of task that those individuals happen to be good at.[31] As the psychologist Earl Hunt has written, "You can find a measure of general intelligence that is *g* and verbally loaded, and produce an advantage for females, or produce a measure of general intelligence that is *g* and loaded on spatial-visual reasoning, and find an advantage for males."[32]

As we saw with height, the correlation of some measure of *g* with economic rewards can serve as a justification for those economic rewards only if the strength of that correlation is not exaggerated either by irrelevant inclusion of some extraneous characteristic that happens to favor one sex's test performance or by the unjustified exclusion of some genuine component of talent that is also underrewarded by economic performance. Most reputable general psychometric tests on which comparisons are based do not suffer from blatant examples of irrelevant inclusion (as they would if, for example, the ability to remember football scores were one of the component tests). But unjustified exclusion is another matter. A test battery that is verbally loaded could be considered to be unjustifiably excluding a number of spatial-visual tests, while a test battery that is visually-spatially loaded could be considered to be unjustifiably excluding a number of verbal tests. Both kinds of battery may exclude other dimensions of genuine talent along which men and women may differ. Observing the correlation of the resulting *g* measure with economic rewards could not tell you whether those rewards "really" reflected talent.

Notice how the argument with respect to gender is quite different from the argument with respect to height. The test batteries do not consist of one group of tests on which tall people do better on average and another group on which short people do better: tall people do better, on average, on all of them. So while you can argue whether the tests exaggerate the cognitive disadvantages of short people, and whether those disadvantages are inherited or acquired, it's not easy to claim that they arbitrarily exclude equally reasonable component tests on which short people would perform better. And different weightings of the component tests might affect the size of the performance difference between short people and tall people, but it would not affect the fact that such a difference existed. This shows, though, how challenging it would be to demonstrate that differences in talent were the real explanation for gender differences in economic rewards. The fact that the exercise has been done, carefully and broadly convincingly, for height only underlines the fact that it has not been done for gender.

Unlike in the case of height, when you have some tests on which men do better and others on which women do better, you can't use the "average" effect across the two types of test to justify economic rewards unless you have separately found a way to justify the weighting of the different components in the average. This isn't some pedantic technical objection: on the contrary, we face this kind of difficulty all the time in making comparisons between complex, multidimensional options. You can't choose between a fun but insecure job and a secure but dull one by ranking each job according to an index that gives arbitrary weights to security and to fun: you really have to decide how much security and fun are important to you. You can't settle an argument about whether it's better to live in the city or the country—even if you thought that question had an objective answer—just by comparing places to live on an index that includes an arbitrary mix of dimensions that tend to favor cities (like quality of nightclubs) with others that tend to favor the country (like the extent of fields and woodland).[33] Even if each dimension captures something we all agree is important, how to weigh the dimensions against each other can never be decided by a purely technical procedure.

It's uncontroversial that there are many aspects of talent that the psychometric tests aiming to measure intelligence cannot capture. That's not because no one has considered the problem but because the very basis of psychometric testing (getting subjects to sit down with a pen and paper or a keyboard and mouse) is like capturing a moving scene in a pencil drawing: the medium intrinsically misses some important features of what is going on. For instance, it doesn't capture reactions between people, a sensitivity to which is one of the most important talents that anybody can exercise but which intelligence tests are by their nature ill-equipped to assess. Nor, usually, does it capture spatial orientation, the ability to find your way around, which is related to certain types of visual spatial ability but distinct from them (though various kinds of computer simulation make it easier to test this than it used to be).[34]

Another important talent that tests usually fail to assess is the ability to form accurate memories of a scene and to use them to inform our dealings with other people. One of the most famous case histories in psychology concerns a patient known as H.M., who, because of the removal of part of his brain to control severe epileptic seizures, was unable to form short-term memories and as a result was entirely incapable of functioning socially in a normal way. Yet H.M. scored high on several standard intelligence tests.[35] This is not a criticism of intelligence tests: it's a criticism of people who think that intelligence tests tell you all you need to know about the value of someone's economic contribution to society. Similarly, given what we know about the multidimensional nature of talent and the accumulated evidence that men and women tend to cluster differently along different dimensions of that talent, if we were to discover that some purported measure of general intelligence conclusively showed either sex to be superior to the other, that would tell us a lot about the implicit weighting of the measure and little about the men and women it was supposed to be measuring.

Personality and Talent

Talent is not just about cognitive skills but also about certain other types of ability, such as hard work and organization. Might gender differences in these noncognitive abilities be more systematic than differences in cognitive skills? A reasonable consensus in psychology has identified five major personality traits (the "Big Five") that are generally heritable and reasonably stable over time (though less so than cognitive skills).[36] They are also found fairly consistently across cultures, though not entirely so (for instance, richer countries display larger variations in reported personality traits).[37] These five traits are openness to experience, conscientiousness, agreeableness, emotional stability, and extraversion. Unlike cognitive skills, they are measured by tests that depend on honest self-reporting, which is why they are much more rarely used for recruitment purposes than are IQ tests: people applying for jobs in marketing know better than to give answers indicating low scores on the extraversion scale. This may also make them more dependent than IQ tests on norms about the kinds of personality it is desirable to have. When it's considered cool to be open to experience, you can bet many more people will give answers that claim to be just that. (This doesn't mean that the answers are simply false: being open to experience really is easier when people around you encourage you in being so.) Indeed, the accuracy of these tests may even depend on the kinds of personalities who take them (conscientious people may be much more cautious in making claims about their conscientiousness). We need, therefore, to be careful before we grant too much authority to any particular personality test.

All the same, psychologists have found gender differences in some of the Big Five traits. Economists have found that such traits also have a tendency to predict labor-market performance, although the correlations are all smaller than those of IQ tests, are less uniformly corroborated across different studies, and are not always either statistically significant or economically very important.[38] Conscientiousness is the trait most strongly (and positively) associated with labor-market outcomes (as Woody Allen is credited with observing, "Eighty percent of success is showing up"), but it is only about half as predictive as IQ. Most (though not all) studies find that women score higher on conscientiousness.[39] Emotional stability, on which men score consistently higher on average, is also positively associated with labor-market outcomes, but to a lesser degree.[40] Agreeableness is negatively associated with labor-market outcomes, but mainly for men. Openness to experience and extraversion are only weakly associated with higher performance, and there are few consistent findings across the many different studies.[41] Furthermore, the most careful recent study to examine the possible causal channels by which personality might affect labor-market outcomes argues that much of the effect happens through its influence on educational achievement.[42] There is certainly evidence that personality affects educational achievement,[43] but the gender differences in labor-market performance that we are interested in are large even when educational achievement is taken into account.

Overall, although it seems a reasonable guess that gender differences in noncognitive abilities *could* have a stronger association with talent than do measures of IQ, none of the studies conducted so far suggests that they actually explain more than a small fraction of the difference in labor-market outcomes. This is partly because the association between personality and labor-market performance is typically weaker and less consistent across studies than it is for IQ; it is also because the gender differences in personality traits are not systematically favorable to men. Men score higher on emotional stability but typically lower on conscientiousness; they are also penalized for agreeableness to an extent that women are not. These differences therefore tend to cancel out in terms of overall average effects. One recent study concludes that "only 3 to 4 percent of the gender gap is explained by differences in personality including differences in traits and trait returns," and no statistical study has made a credible claim that personality differences explain most of the gap.[44] Overall, then, average differences in talent between men and women do not seem any more promising an explanation of the gender gap in earnings when personality differences are taken into account than when we consider only cognitive skills.

Are Men More Extreme?

It has sometimes been claimed that even if mean scores on psychometric tests show very small differences, if any, between men and women, men's scores show a tendency to higher variance, which would explain why men might be more highly represented at the top extreme of the distribution of economic rewards (as well as at the bottom). This tendency to higher variance has also been given a genetic explanation, on the grounds that for psychological traits influenced by genes on the X chromosome, the fact that men carry only one X chromosome, while women carry two, means that men will be more vulnerable to the effect of unusual alleles (which in women would more commonly be offset by the "normal" allele on the other chromosome).[45]

There is indeed evidence for several of the building blocks of this argument. The X chromosome appears to carry a particularly large number of genes that are involved in cognitive development (a fact that incidentally tends to support Geoffrey Miller's hypothesis that sexual selection by women was an important factor in human cognitive evolution, since the X chromosome also carries a particularly high frequency of genes related to sex and reproduction).[46] Measures of *g* tend to show somewhat higher variance for men, and small differences in variance can translate into large differences in representation at the extremes of the distribution of test scores.[47] Scores on tests of more specialized abilities can show even greater differences in representation at the extremes (although, intriguingly, some such discrepancies have declined strikingly over time, suggesting a strong influence of socialization).[48] There are, for example, many more male than female students in the extreme upper tail of the distribution of scores on various advanced mathematical tests, though there remains much controversy about the likely explanation for the gap.[49]

Nevertheless, there's a serious flaw in the argument that differences in variance are likely to account for differences in representation of men and women at the upper extreme of the distribution of economic outcomes. The flaw lies in the implicit assumption that these traits, as measured by test scores, translate directly into economic outcomes, so that more of the trait leads on average to a higher outcome, however strongly the trait is present. That may be true at the lower end of the distribution, since single genetic mutations can interfere in important ways in normal cognitive development, and the incidence of mental disability is around 30 percent higher in men than in women.[50] But within the normal range as well as at the upper end of the distribution, traits determine economic outcomes, to the extent that they do, in the context of complex interactions between the bearer of the trait and other individuals. People who possess single traits to an extreme degree do not necessarily enjoy extreme rewards as a result. Think, for instance, of a trait like talkativeness. Talkative people tend to be economically more successful than very taciturn people. But this doesn't mean that extremely talkative people are extremely successful: on the contrary, unstoppable windbags tend to annoy others to a degree that can seriously hinder their professional prospects. So an argument that men are more likely to have extreme traits, even if true, doesn't tell us anything about whether these traits are the reason for their presence at the extreme of the distribution of economic rewards.

In fact, the kinds of ability that psychometric tests can uncover are very rarely associated unambiguously with success in any field (other than the field of taking psychometric tests). For example, visual and spatial ability, on which women tend to score less well on average than men, is certainly relevant for driving skills. Yet most large-scale statistical studies find that women are better drivers, on average, than men.[51] Visual and spatial skills are important, certainly, but so are risk assessment and courtesy to other road users. Even if individuals who score highly on visual and spatial skills might drive better, on the whole, than those who score less highly, the effect might not go all the way: those with extremely high visual and spatial skills might overestimate their ability to handle the risks posed by other drivers with lower levels of skill or become impatient at having to share the road with them.

Similar points, incidentally, can be made about the association between underlying neurophysiological influences on certain skills and the skills themselves: for example, increases in testosterone appear to enhance spatial reasoning abilities in women and in men with low testosterone levels. But they *decrease* spatial reasoning abilities in men with normal or high testosterone levels, a finding that will surprise no one who has spent time navigating kitchen space in the company of teenage boys.[52] As it happens, men have higher average testosterone levels than women and score higher on certain visual-spatial tests, and the former phenomenon may well be causally responsible for the latter. But because the effect of testosterone on skills does not continue to increase beyond a certain level, if men and women had the same average testosterone levels while men had higher variance, men's overall scores on visual-spatial tests would tend to be lower, not higher, than those of women. This illustrates a general point: the theory that men are more extreme because they have only one X chromosome appeals to a simple model in which genes have a continuously increasing influence on physical traits like testosterone, physical traits have a continuously increasing influence on behavioral traits, and behavioral traits have a continuously increasing influence on economic outcomes. The model is seductively easy to think about, but in lots of ways it doesn't fit the evidence.

Overall, therefore, the claim that the major differences in economic rewards of men and women can be explained by differences in talent—either on average or specifically at the upper extreme of the distribution of talent—is entirely unconvincing. We know that talent can produce higher test scores, and we know that higher test scores are correlated with economic rewards. But outside a small number of fairly idiosyncratic occupations, we don't know that the particular tests on which men perform better than women uncover the particular talents that make people do their jobs better overall.[53] Better visual-spatial skills might help you negotiate the water cooler with more dexterity, or maybe do better on the golf course, but these differences are a slim basis for an overall theory of what makes people in a modern economy do their jobs well.

It's not impossible that someone might come up with such a theory in the future, a theory that could really explain why those aspects of talent on which the test scores favor men are also the aspects that count for most in a modern economy. So none of what I've written here should be interpreted as ruling out the possibility of a talent-based explanation. But no one, to my knowledge, has come up with one yet. In the absence of such an account, particular tests of talent cannot be justified on the grounds that the scores they yield are correlated with economic rewards when those economic rewards are in turn justified because of their correlation with individuals' scores.

If it's not talent that explains the discrepancy in economic rewards between women and men (and if, as we saw, it's not height either), what *is* the explanation? Chapter 6 looks at a different possibility: instead of differences in talent, could the explanation be that men and women have different tastes?

SIX

What Do Women Want?

The great question that has never been answered, and which I have not yet been able to answer, despite my thirty years of research into the feminine soul, is "What does a woman want?"

—Sigmund Freud, letter to Marie Bonaparte, 1926

Different Preferences

IF WOMEN ARE NO LESS TALENTED than men, why don't they receive similar rewards? Here are three possible explanations. The first appeals to the idea that women might simply have, on average, different preferences from men: even if they could do any job just as well as men, they might not necessarily want to, and the jobs they want to do might be less well rewarded whoever did them. Within any occupation they might also avoid some of the positions that happen to be most highly rewarded (for instance, they might be more averse to risk or to aggressive competition). There is some evidence in support of this view, though it's far from conclusive.[1] The other two explanations both appeal to the idea that women get a smaller share than men of the value their work creates. It might be that, for a variety of reasons, women negotiate less aggressively with employers and so end up with a smaller share than men of the benefits that their work creates for their employers. There is evidence in favor of this possibility as well, though it is sketchy and its interpretation is controversial. Alternatively, employers might negotiate more aggressively with women, either from explicit prejudice or because of norms and customs, explicit or implicit, that are a hangover from earlier times. There's a large amount of anecdotal evidence in favor of this third possibility, but again, it's not easy to judge how scientifically well-founded it is or how much of the discrepancy in economic rewards it can explain. This third mechanism may interact with the first: women may have somewhat different preferences from men, but aggressive negotiation by employers may lead to women's paying an unreasonably high price for those different preferences.[2]

The idea that, on average, men and women might have different preferences that influence their professional choices should not really be considered controversial, but it is. Consider, for example, some of the professions where women are systematically overrepresented, such as psychotherapy. It is hard to believe that hidden barriers keep men out of careers in psychotherapy: indeed, there are frequent reports of the vain efforts that are made by psychotherapy educators to attract more male candidates.[3] It doesn't follow that whatever preferences make men less likely to be attracted by psychotherapy are innate (though they could be); they could have been formed by socialization in childhood or may simply reflect perceived preferences of others. But whatever their origins, different preferences of adult women and men, on average, for the professional activities of psychotherapists remains much the most likely explanation for the discrepancy in their representation. If that's true for some professions where women are overrepresented, it may also be true for some in which they are underrepresented: airline pilots may be one example, jazz drummers another. If any reader can point me to a scientific study that finds as many girls as boys dreaming in kindergarten of becoming airline pilots when they grow up, I shall gladly recant.

Still, it would be good to have harder evidence for the existence of such differences in preference. The scientific evidence that we do have comes largely from laboratory experiments. It's controversial how fully behavior in the laboratory can explain the way people behave in life outside the laboratory, but nobody seriously thinks that's because subjects in laboratories behave differently from the way they act in normal life. In one sense, of course they behave differently: that's the point of the laboratory setting. In a laboratory, the experimenter aims to screen out other confounding influences so as to focus on the dimension of interest—in this case, some possible preference difference. By making two choices identical in all respects, except that one is riskier than another and has a higher expected payoff, it's possible to see whether the subjects find the trade-off acceptable—and whether women, for instance, are less likely than men to do so.

There's a lot to criticize in laboratory studies—for example, the common tendency to use university students in prosperous countries and generalize from them to the whole population. And the very structured setting may induce kinds of behavior that you wouldn't find elsewhere—though experimental laboratories are not intrinsically more artificial than other kinds of setting people routinely think of as part of the "real" world, such as stock exchanges and company boardrooms. So although we should keep an open mind and never think of laboratory behavior as an infallible mirror of behavior in other settings, when we find systematic differences between how men and women behave in the laboratory, we should be prepared to look hard for similar differences in their behavior outside the lab.

Two kinds of laboratory experiment have been particularly influential in shedding light on women's preferences. The first consists of experiments showing that women tend to be more risk-averse than men.[4] The fact that these findings have been replicated many times suggests they are capturing something systematic about human behavior. (There is also a related but distinct set of findings that men tend to be more overconfident than women about their ability and likely future performance, which leads to what looks very like risk-seeking behavior.)[5] There are good reasons to expect females of any species to be more averse to risk than males. The difference in reproductive success between the most successful and the least successful females is much smaller than the difference between the most and the least successful males; consequently, natural selection has favored more risk-taking strategies on the part of males.[6] And studies of risk taking in many other animals confirm this conjecture.[7]

It seems plausible that if men are willing on average to take more risk, those for whom the risk pays off would end up systematically better rewarded than women of equal ability. They might also prosper because of promotion procedures within firms that are unable to distinguish between talent and luck.[8] If success is partly due to talent and partly due to good luck, individuals who take bigger risks could, when those risks pay off, appear to have greater talent. (Similarly, when the risks fail to pay off, they will appear to have less talent, which might explain some of the overrepresentation of men at the bottom end of the distribution of society's rewards.) Thus fallible promotion systems may erroneously identify men as having greater talent when in fact they have just had better luck, amplified by their greater willingness to trust to luck.[9] In fact, the possibility that certain reward systems may encourage excessive risk taking (a possibility that the recent financial crisis has given new plausibility) may mean that in such situations, men may receive higher economic rewards on average when the value they create for others is *lower* than that created by women. This may happen in a range of ways and not just because of a difference in attitudes toward risk per se: for instance, there is evidence that, depending on the context, women may be more inclined to altruistic behavior than men, and therefore more likely to take into account potentially harmful side effects of their actions.[10] There is also some suggestive cross-country evidence that female government officials are somewhat less prone to corruption than males in the same position, a form of socially valuable behavior for which they remain sadly underrewarded.[11]

The second kind of experimental finding consists of differences between men and women in preferences for competition. One much-cited study consisted of a test of skill in which subjects could choose whether to be rewarded according to their performance relative to other players (a tournament) or according to their absolute performance (a piece-rate scheme). Nearly three-quarters of men but only a third of women chose the tournament scheme, although there was no difference in performance between men and women under either scheme.[12] Some other studies have shown a tendency for women to perform less well under competitive reward schemes than under piece-rate schemes, while men's performance tends to improve under competition. If rewards in the economy at large favor those who are attracted by competitive contexts and whose performance flourishes in such contexts, this might account for the kinds of difference in economic performance we have seen.

Still, there are two main reasons to be wary of such a conclusion. First, as I emphasize in chapter 4, it is simply not true that modern economic life is all about competing. It's as much about managing cooperation as about competition, and no one has yet come up with even a theory, let alone with evidence, as to which kinds of preference are likely to be most valuable to participants in the modern economy. It's a reasonable guess that a preference for competitive environments would be an advantage to people hoping to make it to board level in a Fortune 500 company. But it remains a guess, and we really don't know to what extent a preference for less competitive environments could be considered a handicap. After all, if you were thinking of employing someone with whom you would have to work closely afterward, would you necessarily want to pick the most competitive person you could find? Other experiments suggest that women have a positive preference for teamwork, and there are surely many contexts in a modern economy where this ought to give them a significant advantage in recruitment.[13]

The second reason for caution is that the experimental findings about gender differences in preferences are quite sensitive to context. For instance, women's performance tends to weaken when they compete against men, not when they compete against other women.[14] Women's preference for teamwork seems to depend in part on their greater optimism than men about the quality of their fellow team members, an optimism that is likely to fluctuate according to what else they may know about them.[15] This sensitivity to context is a general feature of experimental studies of both risk aversion and attitudes to competition. In some laboratory settings the differences in risk aversion are large, while in others they are negligible; this variation makes it difficult to know just how important such differences may be in life outside the laboratory. In some experiments it makes an important difference to women's performance whether they are competing against men, while in others it does not.[16] Similarly, results for adults are often not replicated when the experimental subjects are children, contrary to what one might expect if they were the result of stable underlying preferences.[17] Results found in one country are often not found in another.[18] There's also some evidence that differences in preferences are more likely to be found when the tasks on which subjects compete are considered to be typically male tasks (this is an instance of stereotype threat).[19]

Nevertheless, it would be a mistake to dismiss the findings as due solely to context. There is strong evidence, for instance, that genetic or hormonal factors underlie at least some such gender differences, though others have not been convincingly associated with such factors.[20] Overall, it seems safe to say that these differences are intriguing but that their importance in accounting for gender differences in economic rewards remains a matter of speculation. Rigorous studies of this question are still in their infancy. So far, the measured impact of such preferences on labor-market outcomes, like the measured impact of personality differences more generally, has been small.[21]

Women Don't Ask?

A second possible explanation for differences in women's economic rewards is that, even if women create as much economic value for their colleagues and employers as men do, women may negotiate less aggressively for their share of that economic value and so may end up with a smaller reward. *Women Don't Ask* is the title of a deservedly famous book by Linda Babcock and Sara Laschever that presents large numbers of case studies suggesting that an unwillingness to negotiate may cost women a lot: "Negotiating your starting salary for your first job can produce a gain of more than a half-million dollars by the end of your career," they write.[22] In later studies with other colleagues, Babcock has shown that women's attitude toward negotiation is itself very dependent on context: for instance, women negotiate more aggressively when doing so on behalf of others, when the negotiation setting provides cues as to what is a "reasonable" agreement, and when it is framed as an opportunity for nonconfrontational "asking" rather than negotiation.[23] She has also shown that women's reluctance to negotiate aggressively may be based not just on inhibition but also on a shrewd assessment of how their behavior will be viewed by others. In an experiment where individuals were asked to evaluate others as potential colleagues, men gave significantly lower evaluations to women who negotiated rather than simply accepting offers, while female evaluators showed no such tendency.[24]

Like the findings of different preferences for risk and competitive environments, these findings are both plausible and intriguing. But before we accept them as an important factor in explaining gender differences in economic rewards, we need to consider a puzzle. The more convincing the "women don't ask" explanation is in accounting for the lower salaries of women, the less convincing it is as an explanation for the lower representation of women in certain occupations. Someone who's willing to accept a smaller share of the cake as the price of her participation is obviously a more appealing colleague than someone who negotiates hard for a larger share. If it's true that women create as much value for their employers as men do but cost the employers less, the employers would be insane not to prefer employing women.[25] To put it another way, "women don't ask" might be a complete and convincing explanation for low female salaries coupled with female *over*representation in those jobs that pay them low salaries. But it won't suffice as an explanation of low female salaries coupled with female *under*-representation. There must be another reason why employers are not rushing to take advantage of the profit opportunities that these reluctant negotiators are pushing their way.

A Masculine Bubble?

What other explanations are open to us, in the light of what we've seen so far? If women are not, on average, contributing less than men to the environments they work in, either through having less talent than men or through a relative distaste for the environments in which large contributions are to be made, and if the main reason for the discrepancy does not lie in their unwillingness to negotiate as aggressively as men for their share of the gains, only one possibility remains. It is that men, individually or collectively, are negotiating harder with women than they are with other men, whether through prejudice, habit, or conscious coordination.

We saw in chapter 4 that women might be disadvantaged in the distribution of economic rewards because men monopolize a scarce resource, even if women are at least as talented as men in the areas in which their talents can be compared. Is it imaginable that something similar may be happening in a modern economy in which hunting and gathering are no more, in which nothing prevents women from specializing in the production of a scarce resource? In a modern economy, unlike (perhaps) a small hunter-gatherer community, men don't negotiate as a bloc with women as a bloc. Instead, individual employers negotiate with workers, or perhaps with trade unions. There may be pockets of trade-union solidarity or employer coordination, but there exist no mechanisms whereby men as a whole can dictate terms to women as a whole. More men than women represent employers, certainly, but most men are employees, on the other side of the salary bargain. The gender gap in salaries shows up at every level, between male and female employees as well as between female employees and male employers. It's hard to see how men could be colluding to negotiate more aggressively with women, because it's not clear what could be the mechanism that coordinates this collusion.

And without a coordinating mechanism, it's hard in turn to see how the result of millions of individual bilateral negotiations could result in a systematic discrimination against women, or indeed against any other disfavored group. Let's return to the impact of height on earnings and ask whether that could be the result of an irrational prejudice. It seems unlikely, someone might argue, that there could be systematic discrimination in employment and earnings against short people. Whatever Marxists might once have argued, employers are not a single-minded class who collaborate to enforce their will on a reluctant proletariat. There are millions of individual employers with their own idiosyncratic preferences, and if any employers choose to indulge their prejudices by irrationally refusing to employ talented short people except at lower wages than tall people, other employers will soon discover that they can make more money by doing the opposite. It's conceivable that all employers might become infected by the same irrational prejudice, but it would need to be explained how that was possible. A much more likely story is that if short people are less favored by employers, this is because they happen to have, on average, fewer of the various talents that employers need. There won't necessarily be any one single type of talent, and employers won't all be looking for the same ones. But it won't do any employer any good to base hiring decisions on height if height bears no relation to the talents employers really need. So, unless some good reasons can be found for thinking that economic outcomes are systematically biased against short people in a way that would harm the self-interest of the employers themselves, the fact that tests show that tall people do better would be a sign of the reliability of the tests. Or so the argument would go.

This line of argument now looks less convincing than it might once have, particularly in the aftermath of a financial crisis in which it has become clear that irrational prejudices (about the tendency of real-estate prices to go on rising for ever in real terms, for instance) can come to be very widely entrenched even though it would be possible to prosper by repudiating them. Still, the prejudices have to be strongly salient, the kind that get reinforced as people talk about them, and of which there are enough cases in everyone's experience for the prejudice and the anecdote to override the evidence and the science. Many beliefs about health are like this: belief in the efficacy of homeopathic medicine, for instance, which flourishes in the absence of even a shred of rigorous evidence, reinforced by the strong power of placebo effects (for which we still have very little scientific explanation).[26] Maybe a belief in the economic efficacy of height could be just such a prejudice, reinforced in turn by placebo effects (for example, if confidence is good for economic performance, and tall people are more confident because they have come to believe that height is right, they will perform better).

As we saw above, the link between height and economic rewards does seem to be based partly, though only partly, on a genuine connection between height and talent. But we've found no evidence at all for a link between gender and talent, certainly not for the multiple and flexible talents that a modern economy demands. Might prejudices about the existence of such a link nevertheless persist in the labor market, like the ineradicable faith in homeopathy? Could it be a kind of masculine bubble that inflates the earnings of men relative to those of women, even though nobody has any evidence that it makes economic sense?

The evidence from detailed studies of gender and earnings gives us a few tantalizing clues. First of all, given women's lower earnings, you might expect to find that the returns on higher education are lower for women than for men. After all, a college degree costs as much for women as for men, and if they earn less afterward, it sounds like a less profitable investment. In fact, most studies find a *higher* return on schooling for women, which is puzzling. The most likely explanation is that more highly educated women do face some earnings disadvantage relative to equally talented men, but this disadvantage is much less than that faced by less highly educated women.[27] So women who invest in education not only improve their skills in a way that is valuable to employers; they also equip themselves with what they need to get a larger share of the value they create for those employers (including, Babcock and Laschever might say, a greater willingness to ask). This makes sense if there's an element of shared prejudice in the disadvantage; the more educated the women against whom the prejudice is exercised, the harder it is for the prejudice to survive.

A second intriguing clue comes from a detailed study of the earnings of MBAs from a top business school, carried out by Marianne Bertrand, Claudia Goldin, and Lawrence Katz.[28] They show that although the earnings of men and women after they graduate are fairly similar, they soon diverge, with women earning progressively less than men over time. For instance, men earn, on average, $130,000 at graduation compared to $115,000 for women: while nine years later they are earning $400,000 compared to $250,000 for women. Part of the discrepancy appears to be associated with the fact that the men in the study sample had more training than the women before receiving the MBA. But a large part of the story is that women have more career interruptions and work fewer hours, on average, than the men: for instance, after nine years, 30 percent of women have had at least one spell of not working, compared to less than 10 percent of men. These factors are associated with motherhood: women without children do not have more career interruptions than men, do not work fewer hours, and do not suffer a decline in salary relative to men. The authors are careful to check whether there might be any difference in talent between the women who have children and the women who don't, but they find that women who have children in fact tend to have had slightly higher earnings beforehand than women who don't.[29] The outlines of the story are unsurprising: what's interesting about the detail is the very high price that the women pay for these career interruptions. Or rather, it's the price everyone pays for career interruptions: those few men who have career interruptions suffer as great a salary penalty from them as the women do.

The business world analyzed by Bertrand, Goldin, and Katz is not a microcosm of the economy at large. It is a world of the socially and educationally privileged, in which both men and women are trained to negotiate hard for what they can get—and they get a lot. So we must be careful before drawing wider conclusions from the study. Nevertheless, it represents a slice of the US labor market for which concerns about discrimination have often been voiced (the fact that only 2.4 percent of CEOs of Fortune 500 companies are women being a particularly striking example). For this microcosm, then, the picture seems reasonably clear. The different salary dynamics do not appear to be a result of discrimination against women as such: men and women with identical qualifications who make identical decisions about career interruptions do equally well. As the authors put it: "The data do not indicate that MBA women lose more than MBA men for taking time out. It appears that everyone is penalized heavily for deviating from the norm."[30] Instead, the story is one in which the rules for getting ahead emphasize long hours of work, single-minded devotion to the job, and a refusal ever to take a break. MBAs (both men and women) work, on average, sixty hours per week after graduation, with investment bankers averaging a startling seventy-four hours per week. The fact that fewer women want to play by these rules puts women at a disadvantage. In particular, taking career breaks is costly not just at the time but apparently for years and even decades afterward. It's a matter of different preferences, certainly, but also a matter of the high price those preferences entail. And this in turn raises the question of whether these rules make sense in the modern world.

Are the extreme demands made on the participants in the modern workplace really appropriate to the kinds of tasks the twenty-first-century economy requires? The estimates of Bertrand, Goldin, and Katz suggest that people who work fifty-eight hours a week on average are rewarded by a slightly more than proportionate increase in earnings compared to those who work fifty-two hours, even without considering any subsequent differences in prospects for advancement. Are the last hours of their working week really as productive for the employer as that? And if such career breaks underlie the later discrepancies in representation at the top, are the best CEOs really the people who have proved their worth to the company by never doing anything else in their lives but work? Can it really make sense that career breaks taken in someone's twenties and thirties should constrain their opportunities when they are in their fifties and sixties? It's possible that something about modern business organizations really does require such focus and extreme drive. But it's surprising, if so, that the diffusion of information and communication technologies that has been such an important influence of both social and working life, and has driven such a spectacular increase in multitasking in our personal lives, does not appear to be making more of a difference to the necessary single-mindedness of this professional commitment.

We return to these questions in chapter 9, but at this point I want merely to put to rest an argument that is often made in support of such working practices: namely that if they were not efficient, competition would make them disappear. There's a big flaw in this argument, as we saw in chapter 2. As the peacock's tail reminds us, an extraordinary number of highly inefficient social practices survive and spread, both in nature and in human society, because they signal something about those who adopt them. Given the way the signals are interpreted, everyone may have an individual interest in settling on the same equilibrium; but they might all be better off if less wasteful ways to signal their qualities were available. The social codes in which commitment to the goals of the organization can be signaled only by a willingness to work long hours and never take a career break may have outlived their usefulness, but that doesn't mean that individuals can choose to ignore them without penalty.

The Signaling Trap

The modern office is an environment in which mutual signaling and maneuvering are as essential to survival as they were on the African woodland savanna where we first evolved. Busy people go to meetings where they waste their own and everyone else's time purely in order to show how busy they are. Important people spend time pressing the flesh just to remind others of their importance, while unimportant people offer their flesh for pressing in the eternal hope of transcending their unimportance. Senior managers arrive early at work and leave late to signal their diligence to others; junior managers arrive early and leave late in the hope of becoming senior managers who will do exactly the same. Their departure is further delayed by socializing with those whose company weighs on them already, because the prospect is less unattractive than that of imagining their colleagues socializing without them. People who spent their college years broadcasting to others about how much they would love to take time off to work among other people's children in the third world do a sharp about-turn once they become corporate employees: they broadcast no less strenuously their utter antipathy to the idea of taking any time off at all, and certainly not for the benefit of their own children.

Successful business organizations prosper by finding ingenious ways to allow their members to escape mutual signaling and maneuvering just long enough to do the productive work that enables the organization to survive and employ them in the first place. If this smacks of undue cynicism, ask yourself how much of your time yesterday was spent doing things that didn't contribute directly to the value of your work but was directed instead at signaling that value to other people. It's not surprising that decisions about how much time to devote to raising children are riven with anxiety about how these decisions will be decoded by others: some people may pay a large penalty for their decisions because, well, it's always been that way. Changes that might benefit everyone may not be within anyone's individual reach.

It seems likely, too, that the kinds of signaling women feel the need to do in the workplace may be more complex than those required of men and therefore more difficult for their colleagues to interpret. In chapter 2 I mention the idea that many minority groups feel under pressure to engage in "covering"—not exactly hiding their identity, but making it inconspicuous so that it no longer looms large on the radar of colleagues and friends who don't share that particular identity. Kenji Yoshino cites the case of Margaret Thatcher, who trained with a voice coach to lower the timbre of her voice.[31] Any woman who works in a largely male environment recognizes the challenge of finding a way to project herself as competent and professional while not treating her femininity as a shameful secret. This isn't to say that these pressures are unique to women and ethnic minorities: those middle-aged white males in identical dark suits are toning down something too. Everyone has a hinterland of which their professional colleagues rarely glimpse more than a shadow. But worry about looking masculine doesn't oblige men to undersell many of their best talents in the same way that worry about looking feminine may do for their female colleagues.

If women are stuck in a signaling trap, this may explain not only why they pay such a high price for their career breaks but also why a failure to negotiate may be so costly. I suggested above that someone who's willing to accept a smaller share of the cake as the price of her participation is "obviously" a more appealing colleague than someone who negotiates hard for a larger share—but suppose it isn't obvious after all? Suppose that others routinely assume that if she doesn't negotiate hard, it's because she's not worth so much? Then we might understand why lower salaries for women do not lead to overrepresentation in the jobs that are paying them less. So it becomes urgent to see whether such a signaling trap exists, and if so, how it manages to persist when it leads to the persistent undervaluation of women's talents.

Signaling traps may matter outside the workplace too. If women still want to take career breaks to look after children more often than men do, that may not just be because social codes don't yet put enough value on paternity breaks for men. More subtly, women may be more aware than men of the need to signal conscientiousness through the way they take care of children. Signaling maternal qualities might have been particularly important in hunter-gatherer communities where a woman might need to find other sources of support if the father of her children died, and the need to do so may have become psychologically rooted to a degree that no longer reflects the much safer environments in which children grow up. If so, it's an unfortunate historical development that the family and the workplace have now become so separated. Women caring for children signal a quality—conscientiousness—that employers really value, as we saw in chapter 6. But employers are not present to observe them with their children, and women continue to pay a high price for their absence from the workplace during those years.

Social codes survive because they are enforced in networks of people who live by them. Someone may wonder why, if talented women are underrewarded by modern labor markets, intelligent entrepreneurs (male or female) do not spot a profit opportunity and turn those talents to creative use. The best answer may be that it is one thing for such talented women to exist and another thing for intelligent entrepreneurs to identify and deploy them. The reason this may be difficult is that everyone, male or female, lives and works in networks that shape how we present ourselves to the wider world, and talented but underrewarded women may have less visibility in the networks than their better-rewarded male colleagues enjoy. The importance of networks for our social and economic life is central to our heritage as social primates, and it is the subject of chapter 7.

SEVEN

Coalitions of the Willing

I have never made a friend from whom I could not separate, and I have never made an enemy whom I could not approach.

—Tancredo Neves, president-elect of Brazil, 1985

Fighting and Reconciliation

ALL GOOD PRIMATOLOGISTS KNOW how important it is to be alert to the power of alliances, to the shifting currents of loyalty and betrayal that can make or destroy an individual's standing within the group. And that's just when dealing with their fellow primatologists. When they study the fate of individuals within a group of baboons or chimpanzees, they know that strength, cunning, and luck are not enough. Without the ability to win the support of others and to call on that support in the face of unexpected challenges, a group-living primate is a "poor bare, forked animal" alone in the storm.

Frans de Waal has been studying chimpanzees for nearly four decades and has gained worldwide fame for bringing the political maneuverings of chimpanzees to public as well as professional attention. From 1975 to 1981 he and a group of his students made a detailed study of fighting and reconciliation among the chimpanzees at the colony in Arnhem Zoo in the Netherlands. He believes that both aggression and reconciliation are forms of behavior that reflect adaptive strategies by primates, not merely some inexplicable breakdown of an otherwise rational social order. He found some important differences between the behavior of male and female chimpanzees: in particular, although males in a group are much more likely to fight than females, "reconciliation occurs after 47 percent of conflicts among adult males, but after only 18 percent of those among adult females, with reconciliation between the sexes falling in between." This difference forms part of a pattern: "Among males, most cooperation seems of a transactional nature; they help one another on a tit-for-tat basis. Females, in contrast, base their cooperation on kinship and personal preference."[1]

The result is that coalitions of males are both unstable and flexible: they form and dissolve according to the needs of the moment, and breakdown of cooperation at one moment is rarely inimical to its reestablishment a short while later. Female coalitions are more stable and loyal, but although female friends rarely turn on one another, grudges are equally rarely settled, and never just in order to take advantage of some passing foraging opportunity. "Whereas females spring into action mostly to defend their offspring or closest friends, male coalitions are much harder to predict, as males frequently team up against individuals whom they normally prefer as grooming and contact partners."[2] Compared to females, it seems, males are much more willing to reconcile with their enemies and much more willing to betray their friends.

These findings of gender differences for captive chimpanzees have been broadly corroborated for chimpanzees in the wild in the groups studied by Jane Goodall and Toshisada Nishida and their respective research teams.[3] They also appear to hold for rhesus monkeys, among the most aggressive of all primates.[4] Such findings are not surprising: as the evolutionary ecologist Bobbi S. Low puts it in her book *Why Sex Matters*, "Coalitions, like so many other phenomena, can be a reproductive strategy; and if this is true, male and female coalitions will tend to be different."[5] Sure enough, the size, purpose, and duration of male coalitions have been shown to be different from the size, purpose, and duration of female coalitions across a range of primate species, as well as in some nonprimates, like dolphins.[6] In the terminology coined in the 1970s by the sociologist Mark Granovetter, male primates appear to invest in networks with a greater proportion of "weak ties," while female primates invest in networks with a greater proportion of "strong ties."[7]

It's natural to wonder whether such findings have relevance for human beings. Are similar gender differences found in human coalitions, and if so, do they matter? Do coalitions of male *Homo sapiens* have the more flexible and opportunistic character that we see in coalitions of male chimpanzees, and might this explain a tendency for men to navigate their way in modern labor markets more profitably than women, on average, can do? There is a lot of evidence testifying to the existence of differences in the way in which men and women construct networks and coalitions. Unfortunately, there is limited agreement in the literature as to what those differences are and to what extent they are stable and systematic, rather than varying randomly across different professional and social contexts. There's also little agreement as to whether differences between men's and women's networks reflect different preferences, as opposed to differences in the opportunities that men and women have, on average, to meet and spend time with others. If we were to learn, for example, that women don't tend to belong to clubs that admit only CEOs of Fortune 500 companies, that would hardly teach us much about why there are so few women CEOs in the first place. It would be the result of whatever factors explained this discrepancy, rather than an explanation of them.

Most of all, the relevance of these primate studies to human behavior is hard to assess because human beings have a much richer range of interactions than do other primates. True, like primates, we feed, fight, and have sex together (not always in that order). But we can also telephone each other; exchange business cards; employ, blackmail, or torture one another; vote for the same political party; attend the same birthday party; smile or sneer at each other; follow each other on Twitter; denounce one another anonymously to the police; work in the same office; tickle each other; pay money to each other; and live on the same street. Primatologists examining chimpanzee coalitions have a shrewd idea which kinds of behavior to look for (mutual grooming, for instance, is often a good predictor of other kinds of interaction between the individuals concerned). Primatologists examining human coalitions barely know where to start.

Strong and Weak Ties: The Evidence

One promising place to begin is by examining coalitions along two dimensions: first, those that involve real links between individuals, as opposed to simply common membership in a group, and second, those in which the links involve significant investments of time and effort by both parties.[8] The general finding from primate studies is that females tend to invest more than males in links that involve substantial and repeated investments ("strong ties"), while males tend to invest more than females in links that involve lower levels of investment but to have more of these links ("weak ties"). Strong ties can be characterized as close friendships, weak ties as casual friendships or acquaintanceships.

Among women, such a tendency would be an understandable result of a greater selectivity about relationships, as well as a greater degree of investment in those relationships they undertake. Both of these preferences are consistent with (and predicted by) the logic of sexual selection as far as sexual relationships are concerned. What is novel about this evidence is that they appear to influence behavior over a much broader range of ties, both sexual and nonsexual.

The reason that gender differences in the proportions of strong and weak ties might matter for subsequent economic outcomes was first discovered by Mark Granovetter in a study of the importance of people's personal contacts for their ability to find jobs.[9] You might expect that the sort of links that would really help you to find a job would be your strong ties, because these are likely to be the people who like you best and are most committed to helping you. But these people are often very similar to you and tend to know the same kinds of things and people that you do: consequently their advice and contacts, however sincerely offered, are frequently redundant. Weak ties are much more helpful when you look for a job, because these acquaintances' lower commitment to helping you is more than compensated for by the fact that they are likely to hear of opportunities that you don't already know about. In fact, telling people about job opportunities is a prime example of the benefits of personal networks: it costs very little to do someone a favor, but it may bring them a valuable benefit.

This phenomenon is an instance of what economists call *externalities:* each time I invest a small effort in a contact with someone else, I bring myself a potential future benefit, certainly, but I also create a benefit for them because of the possible information I may some day pass their way. This benefit for them is something I don't necessarily take into account when I decide what investments of time and effort to make, and as a result all networks tend to be smaller than they would be if we took such benefits fully into account. In other words, we may not build links as often, or invest as much time and energy in them, as it would be collectively beneficial to do.

If it's true, on average, that women tend to invest less than men in weak ties, partly because of constraints on their ability to do so but also partly because of preferences that are a shared component of our primate heritage, that might account for women's networks being systematically less effective at helping them to find jobs. Not all jobs, of course: you can apply for many jobs through standard procedures (such as by answering an advertisement) without needing to receive any information from people in the know. But for some jobs—and particularly very senior positions in companies—recruitment takes place informally, through word of mouth or head-hunting, and if women have systematically less access to this information than men do, this might account for some of the discrepancy in representation that we noted in chapter 5.

So is there any evidence for this conjecture? Let's begin by looking at evidence that women network differently from men before considering evidence about any effect this may have on professional rewards. A study published by the sociologist Gwen Moore in 1990 looked at the answers given by participants in the US General Social Survey when requested to name up to five people with whom they had discussed "important matters" in the previous six months.[10] Moore noted that women named a higher proportion of kin than men, and men named a higher proportion of coworkers. She then showed that a substantial part of the difference could be accounted for by the different opportunities faced by women, notably in their lower rates of employment. Still, some gender differences remained, notably in that the women cited more kin, and more types of kin, than men in similar employment and other situations.

This is certainly consistent with our hypothesis, though because the focus was on discussion of "important matters," we can't be sure whether men were investing in additional weak ties: indeed, the average numbers of contacts named by men and women were about the same (almost exactly three in each case). The evidence is also consistent with the hypothesis that men have as many strong ties as women but that their strong ties happen to include more coworkers. One or two other studies provide some corroborating support, but overall there is a frustrating lack of direct evidence for the crucial part of our hypothesis, namely the claim that, whatever their investment in strong ties, men tend on average to have more weak ties of the kind that may be useful to them in their careers.[11]

We lack relevant data partly because the most detailed evidence on networks and their role in career development often comes from studies of single organizations, which tend therefore to focus on ties to coworkers and pay less attention to how these contacts fit into individuals' overall networks. The question of how representative such organizations are of the overall world of work is also hard to escape. Such studies have, however, suggested some important qualifications to the simple view that men focus more on weak ties while women focus more on strong ones. The first qualification is that both men and women display a preference (other things being equal) for networking with members of their own sex. Although unsurprising in itself, this preference has an interesting consequence in organizations where women are underrepresented, because networking primarily with their own sex tends to shut women out of networks of power and influence. The result is that while those professional ties that are most instrumentally useful to men are also the ones that coincide with their social ties, women tend to interact with one group of colleagues (largely female) for personal support and a different group (largely male) for professional help, advice, and advancement.[12] This tendency, documented for the United States, has also been corroborated in a study of Chinese managers, suggesting that it is unlikely to be due to purely local and cultural factors.[13]

The result is that when women do seek professional advice, they do so from colleagues with whom their other interactions are relatively few, and this may sometimes be a handicap. If so, it suggests a second qualification to the simple view: it is not necessarily someone's weak ties as such that are helpful professionally but the right balance of weak and strong ties, with strong ties in appropriately strategic points. The work of the sociologist Ronald Burt has shown in particular how women in organizations in which they are a minority do not necessarily lack weak ties; rather, they lack the legitimacy to make effective use of them. Such women therefore depend on what he calls "borrowing the network of a strategic partner";[14] they need one or two strong ties to be able to benefit from the weak ties of others, whereas men's legitimacy in the hierarchy allows them to use weak ties wherever they are found.

The overall picture, therefore, seems to be one in which different preferences of women and men play some role in the formation of personal and professional networks, but so do the different constraints on and opportunities available to them, in ways that are not easy to disentangle. It's useful, then, to look at a different source of evidence for the role of preferences, which comes from measurement of the ways in which men and women communicate. Many surveys have noted that men and women report using the telephone, for example, in different ways, with women tending to hold longer phone conversations on more personal matters and men tending to use the phone for practical purposes, such as making logistical arrangements or undertaking negotiations.[15] But these data rely on self-reporting, which means that they may reflect stereotypical views of expected gender patterns of behavior rather than actual behavior; many fewer studies have examined actual records of telephone use.[16]

Together with Guido Friebel of the University of Frankfurt, I have looked at direct billing evidence (from anonymized billing records) of the use of phones in both private and professional contexts. We wanted to see if this evidence could tell us something about how men and women communicate, since it is through communicating that they build up their networks. The evidence is not conclusive, but it is certainly intriguing.[17] First, we compared the lengths of calls made by men and women in a random sample of subscribers of a mobile phone company in Italy and Greece during a two-year period from 2006 to 2008. Controlling for other factors such as age and income, calls made by women lasted 16 percent longer than calls made by men; women also made fewer calls. The same was true within each age category: women in their twenties made longer calls than men in their twenties, and so on for women and men in other age groups.

To determine whether such differences might reflect different professional and other opportunities open to women, we examined the records of calls directed (at random) to male and female employees of a call center of a large German consumer services company. The male and female employees work under identical conditions. Calls allocated to women lasted 15 percent longer on average than those allocated to men, controlling for other factors. (Again, the differences between men and women are large and statistically highly significant within each age category as well as for the sample as a whole.) We tested whether this difference might reflect less enthusiasm among women for getting on with the job by looking at operations involving sales. It turns out that women make slightly more sales per shift than men, so they appear to be using systematically different communications strategies and are no less effective as employees.

The one thing that prevents us from concluding that the longer calls of women reflect their different preferences is that they could reflect the preferences of the incoming callers (both male and female) for speaking to women. There's no real way to disentangle the two motivations because, after all, every conversation is negotiated between two people. No one, perhaps not even the parties themselves, can be sure to what extent the overall conversation resembles the conversation they would most have preferred. But though this impossibility may be what gives guaranteed employment to psychoanalysts, it may not really matter for our purposes, because the same is true of all conversations in life. So it seems reasonable to take these data as an indication that communication strategies involving women lead to slightly fewer, slightly longer conversations, and that difference may affect the kinds of network links women make as a result. But it would be good to have more evidence, and no doubt research on social networks in the coming years will make this a priority.[18]

Networking and Professional Success

If the evidence for women's networks being different from those of men is suggestive but not conclusive, can we look at the problem from the other end? Can we look for evidence that women's networks, whatever their structure, are less effective at delivering professional benefits? Research on this question has been inconclusive up to now, largely because of the difficulty of measuring networks and of getting data on both networks and career performance for a sufficiently large and representative group of people.[19] In work with Marie Lalanne of the Toulouse School of Economics, I've been analyzing this very question, using information on around 16,000 individuals who are board members or senior executives of American and European companies (only 9 percent of whom are women). This is a large data set by the standards of most research on this topic, but its size comes at a cost: we don't have information about the actual networking activities of all those people, or even about the people who belong to each other's active social networks. What we do have is information about who has had the *opportunity* to network with whom, because they have worked for the same employer in the past.[20] We don't know whether they chose to make anything of this opportunity. But if it is indeed true that women's networks are somehow less effective at bringing them professional benefits than those of men, this fact should be reflected in our data, because women should be making systematically less of their opportunities than men. In particular, we expect to see two things: first, we can calculate, for each person in our data set, the number of currently powerful people with whom that individual has crossed paths in the past (we can call this the IMS, the Index of Movers and Shakers). We expect that after allowing for the influence of such factors as age and education, those individuals with a larger IMS should have higher salaries if the contacts they have been able to make have yielded professional benefits. Second, if our hypothesis about the greater effectiveness of men's networks is correct, we expect that the effect of IMS on salaries should be larger for men than for women.[21]

Our data do indeed show a very clear effect of IMS on people's salaries. It's not simple to measure, because there's a possibility that both the numbers of opportunities a person has had in their past career and their current salary are joint effects of a third cause, namely their talent. If this were true, simply correlating salaries with the IMS might overstate the effect of one on the other. Adjusting our estimates to take account of this possibility is not straightforward (there are various different ways of doing it, none of which is ideal),[22] but the end result is that we find that people with an IMS value of 250 compared to the average of around 150 would have somewhere between a 2 percent and 4 percent higher salary than the average in 2008, depending on which particular estimate you favor. When we look at the other forms of remuneration, such as stock options, the returns to network size are around twice as large. This is a useful return to being well connected, though not a massive one. Being well connected won't make you rich: it just might make you somewhat better off than you are already. And connections seem to help in two ways. First, being well connected helps you learn about job opportunities out there that would suit you better than your current job. Second, it acts to make you a bit more attractive to your current employer and able to negotiate a little more ambitiously on your own behalf.[23]

As soon as we discovered this effect, we wanted to know whether it worked for women as effectively as it worked for men. And the surprise was that there was no apparent difference at all between men and women on this measure. But when we looked more carefully at our data, we noticed something striking. Board members of companies fall into two distinct categories: executives and nonexecutives. Executives run the company and typically work full-time. Nonexecutives show up for board meetings and sometimes for meetings of various subcommittees of the board; they almost always work part-time, although they may and often do sit on the boards of several companies at the same time. Nonexecutives usually earn much less than executives (in our sample, nonexecutive pay was less than 30 percent of executive pay). And it may not surprise you to learn that only 32 percent of women in our sample are executives, whereas 49 percent of men hold executive jobs. Only 6 percent of our executive sample are women. We have no idea whether this is because prejudice is keeping women out of executive positions or because women are more interested in applying for positions with more flexible working conditions. (Most of the women in our sample are over forty-five, so they are less likely to be constrained by young children than more junior women in the organization.) But it's important to take this difference into account when comparing salaries.

Comparing the salaries of nonexecutive women with those of nonexecutive men does not show a large difference, either in average salaries or in the effect of IMS on salaries. Network contacts are very important here: by our best estimate, having an IMS of 250 instead of 150 raises your salary by 8 percent. And the nonexecutive women clearly use their contacts as effectively as the men. This doesn't mean there's no possibility of discrimination in access to such positions: women still make up only 12 percent of our sample. But the salaries of those who hold nonexecutive positions seem (almost) to match those of similarly qualified men. We therefore can't rule out the possibility, discussed in chapter 6, that women's lower representation is due in part to different preferences for such occupations.

With executive board members, it's a different story. In our sample, women executives earn on average 30 percent less than similarly qualified men. When we include data on other forms of remuneration (such as stock options), the difference looks even bigger.[24] But when we look at the effect of contacts on remuneration, we see something striking. Male executives get a benefit from their contacts: those with 250 contacts instead of the average 150 have 3–4 percent higher salaries and around 10 percent higher levels of stock options and other such indirect forms of remuneration. But female executives don't. It's as though all those contacts earlier in their careers were bringing them no benefit at all. The exceptions are a minority of women who happen to have a lot of currently powerful women among their contacts: those contacts turn out to be fairly valuable to them after all. And after you adjust for the generally different productivity of men's and women's contacts in influencing their salaries, the effect of gender as such becomes small and no longer statistically significant. It looks as though our hypothesis is confirmed for executives and rejected for nonexecutives.

To put it another way, male networks seem to be giving men an edge compared to women when it comes to running companies, but when it comes to just sitting on boards, the two sets of networks are more or less on a par. Once again, women are most disadvantaged when it comes to positions of real power. And this suggests, by the way, that even if policies designed to increase women's representation in the boardroom result in substantial increases in the proportions of women among nonexecutive directors, they may make little difference to the proportions of women among executives, particularly among those who wield real power, such as CEOs.[25]

The fact that these results hold only for executives means there can't be some universal difference in behavior between men and women, which operates regardless of context. Whether women's networks bring them professional benefits depends not just on how women behave but also on how men behave in their turn. It's worth remembering, too, that the people we're looking at in this study are a very privileged minority even of business executives, let alone the population as a whole. So we don't know to what extent we can draw general conclusions about the behavior of men and women from the results of studies such as these. But in some ways the results we've found are stronger than we might have expected, because these are people who are very well connected indeed and probably wouldn't have become board members in the first place unless they were pretty good at networking. To return to one of the themes of chapter 5, it's a bit like looking at the effect of height on basketball scores. If you want to test whether being tall makes you a better scorer at basketball, you shouldn't look just at professional basketball players, because they're all very tall already, and indeed have been selected precisely because of how tall they are. It turns out that among such a tall group, variations in basketball prowess are related mainly to other things and hardly at all to the remaining variation in height.[26]

Similarly, we are dealing here with a group of formidably successful women and men, and they are all probably formidable networkers. So we might have expected that looking for variations in networking ability to explain the remaining differences in professional success would be as misguided as looking for variations in height among professional basketball players to explain the remaining differences in basketball scores. But we have found such an effect all the same. All of this suggests that there's a lot of work left to do before we can confidently conclude that men's networks are responsible for the high price that women pay for the different career choices they make across many different occupations in modern society. But given the evidence that men's networks play a strikingly different role from women's networks when it comes to allocating positions of power in leading companies in both America and Europe, networks remain a leading suspect in the wider world as well.

Network Choices: Prudence or Preference?

It might seem natural to ask whether men and women choose their social and professional networks with an eye to what will be useful for them, or whether they just indulge their preferences for the kinds of people with whom they want to interact, as they might indulge their preferences for the kind of furniture that decorates their homes or offices.[27] But this may be a misleading question to ask, and not just because the answer, if there is one, is almost certainly that people have a mixture of both motivations. The question may be misleading because we have good reasons to think that our preferences for the kind of people with whom we interact have been formed by millions of years of evolution to reflect the kinds of people with whom our ancestors found it most prudent to interact. As we saw in chapter 3, our preferences were a kind of shorthand for what prudence would have dictated, a shorthand suitable for a world of imperfect cognitive abilities and imperfect capacity for commitment. We are attracted to people who smile, so we prefer to be around them; but we also trust them more, and the evidence suggests we're right (on average) to do so.

It seems likely, as Low suggests, that male and female preferences for coalitions developed because these coalitions served important reproductive purposes in prehistory.[28] In particular, individuals chose their coalition partners because those were the individuals they could most trust to bring them the reproductive benefits they needed in the conditions of prehistoric hunting and gathering. If men preferred to interact with men, other things being equal, and women with women, such preferences presumably made sense in the light of the strong division of labor we discussed in chapter 4. But this division of labor is no more. Are the preferences that went with it now entirely out of date? Or might it be true that men still prefer to interact with other men at least partly because they understand them better and can judge better how much they can trust each other? And that women prefer to interact with other women for the same reasons? There is scant and conflicting evidence on whether men are better judges than women of male character and whether women are better judges than men of female character, at least in the workplace.[29] Likewise, there is conflicting evidence from laboratory trust-game experiments about whether men and women behave in a more trustworthy fashion when paired with another player of their own gender.[30]

One possibility is that both men and women interpret signals about the behavior of others in the light of rather coarse-grained generalizations about the groups to which they belong, which in the cases of men judging women and women judging men are more likely to involve judgments about gender (possibly stereotypical ones) than when either group is judging individuals of their own sex. Male employers assessing how seriously other men are committed to the company may well use a wider and more sophisticated set of criteria than those same male employers assessing the commitment of women. We return to the finding of Marianne Bertrand, Claudia Goldin, and Lawrence Katz in chapter 6: women pay a high price for those career choices that they are more likely than men to make. Could it be because they, and the men who employ them, are stuck in a signaling trap in which commitment and talent are simply harder for women to signal than for men?

I return to this question in chapter 9, but before doing that we need to broaden the picture. As Charles Darwin was only too aware, the signals and strategies that men and women use to play the game of sexual selection are just a sample of the signals and strategies that all individuals use to navigate the groups and coalitions that determine the fitness of members of a group-living primate species. As if the anxieties that beset our determination to be accepted by our sexual partners were not burden enough, we all have to worry about how our talents will endear us to our potential collaborators in the workplace independently of concerns about gender. You might have thought that in the modern information economy, these kinds of anxiety might be likely to diminish, as matching technologies in everything from online dating to job search make the process of finding professional and personal partners more efficient. Chapter 8 investigates whether this optimistic vision is too good to be true.

EIGHT

The Scarcity of Charm

Je suis soulfre et salpêtre, vous n'ètes que glace.

(I am sulfur and saltpeter, while you are ice.)
—Pierre de Ronsard, 1578

Scribes and Screevers

ONE OF THE SIDELINES of the Brahmin schoolteacher in the Indian village where I did fieldwork as a graduate student in the 1980s was to write letters on behalf of the villagers who could not write for themselves—nearly all of them, as it happened. They were usually bureaucratic letters—to government officials or personnel officers of large companies—but sometimes they were family letters to sons or cousins who had made good in a city somewhere. Sometimes they were letters seeking a marriage for a son or daughter, letters in which the appearance of literacy and prosperity might make all the difference to a young person's fortune and happiness. I never learned how much the schoolteacher charged for his services, but given the dependence of his profitable sideline on a continuing pool of illiterate fellow villagers, I did sometimes wonder about its effect on his motivation as a teacher.

I remembered the schoolteacher many years later when reading in the *New York Times* in 2007 the story of G. P. Sawant, who is a professional letter writer in Mumbai. Sawant also writes letters to all kinds of recipients: he even does pro bono work on behalf of young women who come to the city from the countryside to work as prostitutes. Many are illiterate, and for obvious reasons most do not want to tell their families exactly what work they are doing. So Sawant has not only to stand in as a writer but also to invent a life for them, a story about their struggles and successes in the city that they can feel comfortable transmitting back to their parents and siblings in their home village. It is a noble and poetic vocation in more ways than one, but his livelihood is now under threat, the *Times* reports, from the spread of cell phones: "Calling the village or sending a text message has all but supplanted the practice of dictating intimacies to someone else."[1]

It sounds poignant, this story of a modern technology driving out the humanity and charm of the old. Modern literature is full of touching metaphors of ways of life vanishing because of modern technology, or of old technologies turning slowly into folkloric rituals until their original point is forgotten. Jean Giraudoux writes of the rural France of his childhood: "A shepherdess who beat her two clogs against each other: twenty years ago it was an alarm against wolves, now it's against foxes, and in twenty years its only use will be against weasels."[2] He clearly means his readers to mourn the richer timbre of life in former times.

But something about this particular story of the disappearance of the letter writers doesn't add up. A century and a half ago, Henry Mayhew devoted several pages of his massive study *London Labour and the London Poor* to "screevers, or writers of begging-letters and petitions," whom he described as "a class of whom the public little imagine either the number or turpitude." "Their histories vary as much as their abilities," continued Mayhew: "Generally speaking they have been clerks, teachers, shopmen, reduced gentlemen, or the illegitimate sons of members of the aristocracy; while others, after having received a liberal education, have broken away from parental control, and commenced the 'profession' in early life, and will probably pursue it to their graves."[3] It was not the moral disapproval of respectable folk that made this profession disappear, but it was evidently not the arrival of mobile telephony either. It was universal literacy.

Literacy has made slow progress in India, but there are proportionally many fewer illiterates than there used to be and many more literate people competing to help them. I'd like to think that the Brahmin schoolteacher in my village is now out of work, or at least out of that kind of work. Quite possibly he has now retired, and his children are selling a different skill to their fellow citizens: computer programming. India now has well over two million computer engineers, nearly as many as the United States, and they earn salaries that may be modest by American standards but are beyond what their parents' generation thought possible.[4] Many more young people in schools and colleges up and down the country, seeing what a computer programmer can earn, are dreaming of the time when they too will have earned their qualifications and can make a good living. But many of their parents also dreamed that a university degree would bring guaranteed prosperity and were disappointed when (in a country that invested heavily in its universities and comparatively little in its primary schools) so many people had degrees that university credentials were no longer worth very much. The current enthusiasm for computing qualifications is likely to bring just as much disappointment in its wake. The fact is that skills acquired through hard work bring a sober and honorable return, but no more than that: the only way to make a fortune is to be skilled in ways that others cannot emulate and that are also in demand. One way to succeed is by acquiring a skill before it becomes fashionable, like the first generation of Indian computer engineers. Another is by acquiring a skill that is kept in artificially limited supply by the institutions that offer the training (like qualifications to practice law). And another is by having a skill that is too difficult for most people to copy, however hard they try.

There is justifiable excitement in India, as well as in the world's other poorest regions, such as sub-Saharan Africa, about the potential of information and communications technology to ease the lives of the very poor. As of late 2011, there were well over half a billion mobile phone subscriptions in Africa and more than three-quarters of a billion apiece in India and China.[5] There are now around 6 billion mobile subscriptions in the world.[6] Even allowing for multiple subscriptions, that means a large majority of the world's population now have mobile phones, gadgets that seemed the stuff of science fiction little more than a quarter of a century ago. Many others have the use of a mobile even if they don't have a subscription. Across Africa you can see people seated by the side of the road, selling mobile phone time by the second to those who don't have a phone of their own. And callers in poor

A camel driver using a cell phone in the Nubian desert in Egypt, 2010. © Jon Bower/Loop Images/Corbis.

countries are not just chatting, agreeable and important though that may be. They are using all kinds of services, like mobile banking and medical consultations, services that may save a livelihood or a life.

The Internet has not yet reached as many people as mobile phones have. The International Telecommunication Union (ITU) estimates that China had 111 million Internet subscribers in 2009, India only 15 million. There are many more users, of course: the ITU estimates them at more than a billion in the Asia Pacific region, 105 million in Africa, and around 2.4 billion in the world—still a minority of the population.[7] But it's an astonishing development in just two decades. Even if the Internet demands higher levels of literacy of its users than the mobile phone network, there's no doubt at all that it will change the lives even of the poor. Many people will gain access to a better standard of education than their geographical and economic situation would ever have permitted without it. (Already it seems likely that several million students have learned scientific subjects to a world-class standard thanks to the Khan Academy, whose free classes are available on

YouTube.)[8] Even one literate person in a family can improve the prospects for the whole family by knowing how to find contacts, markets, jobs, or help in an emergency. If mere literacy makes such a difference to the isolation of the poor, think what a fully Internet-connected education can do.

Yet Internet and communication technologies will improve the lives of the world's citizens only because they connect people to each other. Sometimes connections are valuable because they solve coordination problems, bringing together individuals who would otherwise find themselves in unconnected niches, as we saw in chapter 3. If one of the most tragic predicaments you can imagine is a child dying for lack of medical attention, how much worse is it for the child to die while a doctor sits in an empty surgery a few kilometers away because there is no communication between them? But there is a limit to the benefits connections can bring. More exactly, the benefits of matching technologies are limited by the scarcity of attention: not everyone to whom you want to be connected has the time or inclination to be connected to you. In the information economy, attention is the ultimate scarce resource.[9] Many of the talents that natural selection has instilled in us, or that can be acquired by an expensive education if natural selection has not, are weapons for staking claims to the attention of others, weapons whose exercise means reducing the attention available to our rivals. By the logic of all arms races, the greater the number of people who acquire such talents, the less such talents are worth. More advanced communications technologies will reduce the amount of waste resulting from people's idle attention while increasing the amount of waste resulting from people's congested attention.

The history of writing is a long, epic story of the transformation of the magical into the ordinary. When the first scribes mastered cuneiform or hieroglyphic scripts, the knowledge they acquired was so mysterious to laypeople that it usually brought the writers priestly status as well as material rewards. The spread of education has slowly led to the democratization of writing, and the adoption of alphabetic scripts at different times in various cultures (most recently by fans of text messaging in China) has contributed to the process by making education less difficult and expensive to acquire.

As a result, conventions about what you have to do to prove your talent have evolved: merely being literate impresses nobody any more. The rise of written poetry as an art form (as distinct from the kind that was spoken or sung to an audience) was the natural result of writing having spread widely enough for the mere art of putting meaning onto a page to have lost a good part of its original magic.

Charm in the Modern Workplace

We can see the same arms race at work in more modern skills as well. Little more than a century ago, the manufacture of moving pictures must have seemed as magical to its first, privileged spectators as the art of making writing had done five millennia earlier. When the Lumière Brothers held the first public screening of motion pictures at the Grand Café in Paris in 1895, the flickering images on the wall of the café caused a sensation and resulted in an invitation to tour in London, Bombay, New York, and Buenos Aires the following year. Now, far more technically impressive creations languish on YouTube, watched by a few indulgent friends of the creator at most. To the extent that technical sophistication has given us access to more satisfying creations, it represents progress for the benefit of all of us. But if it's easier to make technically accomplished movies today than it has ever been, it is harder than ever before to make a living doing so. Those who succeed have something extra, something that marks them out from their rivals who have the same technical skill. If you want a word for it, that word is *charm*. Charm sounds like a nice, comforting thing, but it's not: almost by definition, not everyone has it, or not when they need it, and for everyone whose charm is working, someone else is left seething and frustrated.

Charm plays a more important role in some professions than in others. One way of seeing it in action is to note how the earnings of people who do the same job can vary substantially according to how effectively they attract the attention of others: colleagues, customers, collaborators. Almost nobody ever does a job that is wholly independent of the contributions of others, and almost everyone depends on others in a fundamental and far-reaching way. This means that to do your job well, you need to be able to persuade others to work with you; and however conscientiously you do your job in other respects, if you can't persuade others to form a team with you, your rewards will be meager and your sense of fulfillment low.

We can see this in the variations over time of salaries in a number of professions where charm is important. Inequality in earnings has been increasing in recent years in the United States and in many other industrialized countries. You might have thought that was just because certain kinds of already well-paid job were being even better paid, relative to the others, than they were before: bankers' pay rose relative to that of bellhops, for instance. But for some occupations, inequality has been rising even among people doing the same job. And there are good reasons to think this is because the rewards of charm—that essential ingredient that not everyone has enough of—have been rising over time in those occupations as well.

Evidence from the US Bureau of Labor Statistics points in this direction. The BLS publishes salary-inequality figures by some very precisely defined occupational categories (more than eight hundred of them, in fact). We need evidence like this. It wouldn't tell us anything if we discovered that salaries for "car workers," for instance, were becoming more unequal, because that category covers too many different types of skill, from the people who make tea in the General Motors canteen to people who design software in its research division. Widening disparities in earnings within that broad group could have been happening just because technology has been improving faster in computer-aided design than in tea making. So what do we learn from these finely graded occupational earnings statistics? In some occupations, there's a much bigger difference between the earnings of the best-paid and worst-paid practitioners than there is in others. For instance, the BLS reports a number of occupations that in 2008 had the same median earnings but very different distributions. Among signal and track switch repairers, those whose earnings were at the tenth percentile of the distribution took home a little more than $16.50 an hour, while those at the ninetieth percentile took home a little more than $30. Among film and video editors, by contrast, those at the tenth percentile took home just under $12 an hour, while those at the ninetieth percentile took home $54—around four and a half times as much.[10] The point isn't that video editors are better paid than signal and track switch repairers—they're not, because at the median they're paid exactly the same (that's the whole point of this comparison). The point is that getting into the top 10 percent gives you a much bigger salary payoff in one occupation than in the other.

There are many possible explanations for this difference, but the role of charm seems one of the most likely. Any job requires a combination of method and flair, of things you can learn by repetition and things you can't, just as playing any musical instrument requires a mixture of perfected technique and a sense of musical interpretation. It seems likely that repairing signals and track switches needs more method and less flair than video editing does. (That's not to say it needs no flair at all: many stubborn repair problems need imagination as well as dogged persistence.) But the benefits to a flair for signals repair are probably occasional and serendipitous, whereas in video editing they're often the whole point. Every video made today has to compete for attention with a gazillion others. The editor who can make the whole creation sing may have just that edge that makes the difference. And a talented editor is not just someone who can make a random set of video material into a better and more attractive whole. It's someone who can attract, and be sought out by, talented scriptwriters, actors, and directors, people whose collaboration raise the odds of success in the first place. The most talented signal repairer is probably not going to be attracting collaborators to anything like the same degree, and the quality of the collaborators is going to matter a lot less anyway.

In any occupation, it's possible that many factors other than the return on charm may influence the distribution of earnings and may affect comparisons between occupations. So it's worth comparing how the distribution of earnings changed in various occupations between 2000 and 2009. (It's difficult to measure spreads by occupation over longer periods than that, because the Bureau of Labor Statistics regularly changes its definitions of occupations to keep up with the changing nature of the modern workforce.) Let's start by looking at two occupations where it's unlikely that charm plays much of a role in the determination of earnings: parking-lot attendants and fast-food chefs. Inequality is measured as the ratio of earnings at the ninetieth percentile to earnings at the tenth percentile. The spread in these two occupations is indeed as low as it gets, with individuals at the ninetieth percentile earning respectively 1.8 and 1.5 times as much as the individuals at the tenth percentile.[11] Nor did it change to any significant degree in the first decade of this century.

We can compare these to two other occupations where it seems unlikely that charm plays much more of a role than it previously did: accountants and paralegals. These are higher-paying occupations and also have somewhat higher earnings spread between the tenth and ninetieth percentiles: 2.8 times for accountants and 2.5 times for paralegals. That's probably due to the fact that more of these individuals work in teams, and individuals with charm are likely to end up working in higher-paid teams for higher-paying clients. The spreads have edged up very slightly over the decade (in 2000 they were 2.6 for accountants and 2.4 for paralegals), but the change is still fairly small.

The really interesting changes come, though, in various occupations in the motion-picture industry. It would be tempting to think that this is an industry in which earnings generally should have become more unequal, because of the way in which the industry has become polarized between very large studios and many small independent filmmakers, and the way in which the growing internationalization of movie distribution makes a few superstars able to command truly world-beating earnings. But this growing inequality characterizes only some of the occupations within the industry. Film editors, actors, and make-up artists have indeed seen substantial rises in earnings spreads over the first decade of the twenty-first century (spreads for editors rose from 3.8 to 4.5, for make-up artists from 4.2 to 5.7, and for actors, from an already very high of 7.4, to 9.3 in just nine years). But a number of other occupations in the same industry saw spreads fall—notably broadcast technicians (from 4.6 to 3.7), camera operators (from 4.5 to 4.0), and projectionists (from 2.7 to 2.1).

What was going on? In these latter, relatively technical occupations, it seems a reasonable guess that technology was slowly taking the charm out of the work. The same sophisticated and user-friendly technology that enables the likes of you and me to make a high-quality video clip and post it onto YouTube also makes it much harder for a really skilled technician to be creative enough to stand out from the crowd. In short, these skills seem to be offering less and less opportunity for the exercise of creativity as more and more people acquire them.

Statistics like these offer more opportunities for speculation, it has to be said, than for drawing hard conclusions. For instance, it's only to be expected that the spread between the tenth and the ninetieth percentiles in the earnings of fashion models has been rising (from 2.8 to 3.3), and a little surprising that the same measure for writers' earnings has been falling slightly (from 4.0 to 3.8)—until you reflect that the writers who earn really big sums are just a tiny fraction of the total, so the ninetieth percentile is hardly going to capture the earnings of those who write bestsellers. Writing is also a famously solitary occupation, at least compared to most of the occupations in a modern economy. The rising spread in the earnings of bailiffs (3.2 to 3.6) is a poignant commentary on how creative that profession had become by the end of the decade.

Finally, we can consider two statistics for occupations in the computer industry suggesting that the same kind of dual phenomenon may have been occurring there as in the movie industry. The spread in earnings of computer programmers rose slightly (from 2.7 to 2.8), while that of technicians working in computer support roles fell (from 3.0 to 2.7). While there is a lot of overlap between the two occupations, there are some broad differences too. Programmers have to think up creative solutions to novel problems, often in teams. Computer support technicians tend to work more on maintenance and debugging, tend regularly to revisit problems they have seen before, and tend to work on their own or in large call centers that allow them little individual autonomy. Like broadcast technicians, for whom the increasing sophistication of the technology takes some of the challenge out of the work, computer support technicians appear to be finding that the technology, while certainly not making them redundant, is increasing the proportion of routine in their work. And perhaps the increasing proportion of routine may be a sign that future innovations could threaten their livelihood even if it has not done so already. Certainly there is a correlation—small but significant—between the change in earnings spreads and both the change in employment and the change in median earnings over that decade.[12] Higher inequality, both across occupations and over time, seems to have been the price Americans paid for being in occupations with above-average prospects for employment and earnings growth.

Sexual Selection Again

It's important not to build too much on these numbers: they're suggestive, not conclusive. If the argument that computer programmers are not the answer to India's poverty problem (because their current high returns are an artifact of their relative scarcity, which will decline over time) doesn't persuade you, these statistics about earnings inequality won't—and shouldn't—make the difference. All kinds of things may be affecting earnings spreads, and the changing role of charm in these occupations is only one of them. But they may serve as a reminder that technology is not making purely routine the task of building the teams that manage production in a modern economy.

Today, producing goods and services is almost always a matter of teamwork. Sometimes the different members of the team can operate entirely at arm's length, doing their part of the job and passing it on to the next person through a market transaction. Even making so relatively simple an object as a shirt can require coordination of a team whose members are located on several continents, but a large part of that coordination can be brought about by the market. More commonly, though, teams of people work together to manage part of the production process, and before it's possible even to decide how to motivate and manage that team, its different members have to agree to work together.

You might have thought that putting together a production team is largely a matter of searching for the right kinds of people to take part in it. You might also expect that this process would be made vastly easier by the technologies so impressively spawned by the information economy, including, of course, Internet searching itself. But in addition to finding the right kinds of people, putting together teams is also a matter of persuading those people that they should cooperate. And the same search technologies that make the first task easier make the second task more difficult. The more easily Prince Charming can find his Cinderella, the more likely it is that instead of blushing winningly at him, she will reply: "I'll see if I can fit you into my schedule; maybe we should try and have lunch sometime." The very technologies that help us to find our ideal partners, whether in our professional or our romantic lives, crowd the attention of those partners and raise the stakes in the contest to find ways to stand out from the crowd, requiring ever more elaborate and unpredictable ways of staking a claim to their attention. No piece of software, no algorithm, no formal training that anyone can copy will ever guarantee that those we want to work with will want to work with us.

This predicament—that more sophisticated matching technologies raise rather than lower the need for charm—may help to explain a puzzling feature of the networks discussed in chapter 7. Suppose it's indeed true that talented women pay an excessively high price for career choices such as taking career breaks and working less than the mind-numbingly long hours apparently standard for men in some kinds of job. If these choices really mean that such women receive lower rewards than similarly talented men, why aren't smart entrepreneurs sniffing out a profit opportunity by linking up with such women? I suggested the answer might be that entrepreneurs might know such women exist but have difficulty finding them, because women who take career breaks also drop out of the networks that entrepreneurs use to locate potential opportunities. Surely, you might think, this is where modern information technology can make a difference. Yet it's precisely because modern technology is so good at matching up people that it can make our selection problem more difficult rather than easier: on the Internet, everyone is Prince Charming, so only Prince Extra Charming will do. And the more efficiently the software works, the more likely is Cinderella to think that accepting Prince Extra Charming may be a less attractive option than waiting for Prince Utterly Charming to come along. And that's just the problem in the dating market. In the modern version of the story, once the matching problem is solved, Cinderella doesn't live happily ever after but takes a career break. Once she tries to return to work, hoping for something a little more elevated than the hearth sweeping with which she started her career, she learns, to her dismay, that everyone has glass slippers nowadays: they're no big deal.

Networks function because they help us navigate in a world crowded with information. The more technology enables us to rationalize and organize our networks, the more information we have to process, and the more, paradoxically, we may rely on our instinctive and intuitive networks to help us feel comfortable about other people we interact with. An employer looking for committed workers who has only three applicants to choose among may be willing to spend quite a lot of time interviewing them and finding out about their individual qualities. An employer who posts an online job ad and receives three thousand applications may resort to all kinds of unscientific rules of thumb to reduce the applicant pool to a manageable size. Employers scared of the consequences of receiving three thousand job applications may resort to various informal methods to avoid even getting to that point, such as asking the people they know to recommend candidates, and so on.

True, the law nowadays uses sophisticated methods to ensure that once women reach the official short list for a position, they do not suffer discrimination. But these methods may be powerless to ensure that women reach the short list that really matters, namely the unofficial one inside the head of the decision maker in today's informationally overloaded world. If women aspire to fully equal participation in the modern economy, they will have to find ways to level up their access to the most fundamental scarce resource of all in the twenty-first-century world—the bottleneck inside other people's brains.

This bottleneck affects men no less than women. Indeed, some have claimed that it affects men far more. Is that true, and does it mean that questions about male power are no longer relevant?

Is There a Crisis of Men?

As I noted in the introduction to part two, it has become fashionable in recent years to talk about a "crisis of men." Discussions of this phenomenon conflate at least two observations. The first is that society's most marginalized groups—the homeless and the prison population in particular—contain far more men than women. This is certainly true and has always been true. It may be a more important phenomenon than any discussed in this book, but it is not evidence for the "myth of male power," to use the title of one bestselling book published a few years ago.[13] The fact that many men are powerless does not preclude the possibility that some other men may be very powerful indeed.

A second trend that is sometimes used as evidence for a "crisis of men" is that women's college enrollment rates in the United States have substantially overtaken those of men, and that as a result women in their twenties and thirties have higher education levels than those of men. This development is certainly not confined to the United States: it's true of many other industrialized countries (with the important exceptions of Germany, Japan, and Korea).[14] But it's not the result of any change in the educational aspirations of men. According to the US Census Bureau, the proportion of US men in 2010 with at least a bachelor's degree is 30.7 percent for men aged between 65 and 69; for men aged 30 to 34 years it's 29.9 percent (a negligible difference, considering the fact that some people graduate after the age of 35). The trend for roughly three men out of ten to graduate from college has remained essentially unchanged for thirty-five years. What has changed dramatically is the level of women's aspirations. Only 23.3 percent of women aged 65 to 69 have a bachelor's degree or higher credential, while for women aged between 30 and 34, that proportion is 38.2 percent.[15] In terms of educational achievement, women have pulled out as far ahead of men as they were formerly behind. That's a remarkable achievement, but in what sense is it a crisis of men?

A safer assessment of this trend is that it will be bad news for some men and good news for others, and will require a potentially stressful readjustment of priorities for everyone in the meantime. It's not unlike other economic relationships, in fact. When China enjoys faster economic growth than the United States, that's mostly good news for the United States, because it allows US workers to sell more goods to China and US consumers to buy more Chinese-produced goods. Although in some respects China is a rival to the United States (for example, in exporting to third countries), its role as an economic collaborator is far more important. It makes no sense to see economic relations as sporting contests, in which one country's success must mean another country's failure. But in the meantime, some people (those who made money dealing with China when it was poor) will have to raise their game, and the necessary adjustments may create stress on both sides.

Education, likewise, is not a sporting contest, in which a larger number of more highly educated women must portend the failure of men. As professional colleagues, as contributors to public life, and as personal friends, more-educated women are good news in principle for everyone. To their potential life partners, more-educated women are certainly good news too—so long as the women themselves don't have unrealistic expectations about those partners. The average American woman is a lot more educated than her mother, but if she thinks that entitles her to choose as her life partner a man who is a lot more educated than her father, she's riding (on average) for a big disappointment.

That disappointment is real, and it is in the marriage and dating stakes that it now seems to be felt most harshly. Marriage rates (among heterosexual couples) have been falling for many years, with fewer than 60 percent of US white adults married in 2010, compared to more than 70 percent in 1970. No doubt this drop partly reflects changing social conventions about the appropriateness of marriage for committed couples, but it probably also reflects a reevaluation of the extent of commitment many individuals feel to their current partner, if indeed they have one. Greater selectivity by women, many of whom feel their educational achievements entitle them to more educated partners than they can currently find, is likely to be part of the explanation. Greater selectivity by the minority of highly educated men, who now have more women to choose among, is probably involved as well. It's unlikely to be a coincidence that marriage rates have fallen fastest among African Americans, the group in which women's educational achievements have pulled furthest ahead of men's in recent years.[16] However, although there is no shortage of corroborating anecdotes, reliable systematic evidence for these conjectures is not easy to find. The economists Betsey Stevenson and Justin Wolfers have claimed that women's happiness has declined in the United States over the last thirty-five years: if true, this decline may indeed reflect disappointment in relationships induced by such educational disparities.[17] Stevenson and Wolfers find that the happiness of less-educated men fell over the period while that of more-educated men rose. But they also find that more-educated women suffered smaller happiness declines than less-educated women, the opposite of what would have been expected if relationship disappointments driven by educational disparities were the main factor behind these trends.[18]

If these speculations are accurate (and they remain speculations, given the state of current evidence), they certainly suggest a crisis in relationships for less-educated men—many of whom have lost their charm, in a word, for their female contemporaries. But if it's a crisis, it's one that affects women as well, at least as long as the expectations of women fail to adjust to the reality of the new educational imbalance. Education is good for improving your access to many valuable things in life, but access to a fixed stock of potential life partners is something it's hard for everyone to have more of at once.

NINE

The Tender War

Et plus le temps nous fait cortège
Et plus le temps nous fait tourment
Mais n'est-ce pas le pire piège
Que vivre en paix pour des amants?
Bien sûr tu pleures un peu moins tôt
Je me déchire un peu plus tard
Nous protégeons moins nos mystères
On laisse moins faire le hasard
On se méfie du fil de l'eau
*Mais c'est toujours la tendre guerre.**

—Jacques Brel, *La chanson des vieux amants*, 1967

Visiting Anthropologists

NO MEMBER OF ANY OTHER SPECIES than our own has ever been appointed to a university position in social anthropology. That's a pity in lots of ways. We are far from being the only species with a talent for the study of animal behavior: many predators, for example, have an exquisitely fine understanding of the behavioral weaknesses and idiosyncrasies of their prey. Instead we rely for our understanding of human society on the insights of social anthropologists (as well as psychologists, sociologists, and economists) who are, if not exactly blinded, at least desensitized to many of the

* The longer time keeps company / With us, the more it tortures us. / The very worst of lovers' traps / Would be to try to live in peace. / Of course, you cry less easily / These days, of course I crack less soon. / Less jealous of our mysteries, / We leave less in the hands of chance. / The current makes us careful now, / But still the tender war goes on.

truly weird aspects of human behavior by the fact that they belong to the very species they study.

So we have to imagine what anthropologists from another species would make of us. To make it easier, let's stick to the apes: it's too hard to imagine what dolphins might make of our landlubberly clumsiness. Chimpanzees would undoubtedly marvel at our ability to congregate in groups without fighting:

> Members of *Homo sapiens* appear not even to notice, let alone react to, what would seem to us intolerable invasions of their bodily boundaries. Many, even most, individuals are capable of accepting the uninvited proximity of others for an entire day without once coming to blows. Our initially promising field study "Violent Conflicts in the Tokyo Subway" had to be abandoned after weeks of patient observation failed to yield a single recorded incident.

Bonobos would surely be amazed by something else:

> It cannot fail to strike the most unbiased observer that there is a startling discrepancy between the vast amounts of time and energy that members of *Homo sapiens* devote to thinking, talking, and agonizing about sex and the microscopic amounts of time and energy they devote to actually engaging in it. Apparently healthy males and sexually receptive females appear capable of spending hours in each other's company without ever making physical contact, though reliable native informers report that neural activity can become highly agitated under such circumstances. It's not as though they have much else to fill their time: foraging for food requires only two or three hours per day, and the rest of the day is spent in elaborate display activities, involving frequent changes of bodily decoration, that *almost never lead to intercourse.* Our initially promising field study "Sexual Intercourse among High-Ranked Business Executives" had to be abandoned after weeks of patient observation when the only incidents worth recording were so infrequent and of such short duration that they initially

Chimpanzees fighting, Liberia. © Clive Bromhall, Oxford Scientific.

escaped the attention of our field researchers, who were themselves having sex at the time.

What gorilla anthropologists would notice most about *Homo sapiens* (apart from the rarity and furtiveness of our polygamy) would probably be the way in which adults' lives are ruled by children:

> Juvenile members of *Homo sapiens* expect to be fed and groomed long after they are physically capable of foraging. Social codes entitle and even expect juveniles to object to all kinds of physical contact that we would consider routine; assaults on them are considered undesirable and even shocking, rather than an occupational hazard of their proximity to adult males. This deference to juveniles appears to be part of a systematic wider pattern in which alpha individuals frequently permit themselves to be pushed around by the betas. Indeed, high-ranking individuals frequently adopt a kind of forced meekness in the face of coordinated hostility from groups of low-rankers who believe them to have abused their high rank, an attitude that apparently serves only to encourage the betas.

Female bonobos neglecting their anthropological fieldwork. © Frans Lanting / Corbis.

As professional anthropologists, these apes would take care to avoid expressing any shock or even moral disquiet at our behavior but would do their best to understand it as an adaptive response to the constraints of our environment. A mixed-species research team might well begin to conjecture that these three aspects of our lives were somehow connected: our low levels of intragroup violence; our much greater concern with thinking and talking about sex than with actually engaging in it; and our organization of society around the needs of children. By pooling their insights, they might come up with a story that went something like this:

> *Homo sapiens* has colonized an evolutionary niche that depends on making large and elaborately cooperative investments in its offspring.[1]
>
> This niche requires *Homo sapiens* to engage in complex webs of cooperative activity, in which just about every important activity is the product of continuous teamwork. In recent millennia this has also led *Homo sapiens* to live in large, densely populated communities to make the most of this cooperative potential.

But the same overcrowding and mutual dependence that make such cooperation possible also create massive temptations for violent and antisocial activity, notably the killing of rivals for scarce economic and sexual resources but including also a vast number of lesser antisocial acts.[2]

Human societies would therefore implode under the weight of their own social proximity and the complexity of their mutual interdependence unless the actions of their members were continuously supervised and controlled. All human beings live in what are, by ape standards, effectively police states, in which the police are everyone else.

In particular, individuals who transgress against the elaborate social codes that govern acceptable behavior provoke not just individual retaliation but also the collective retaliation of coalitions of others. This applies even (and perhaps especially) to high-status individuals, who, because of their access to powerful means of coercion, would be especially tempted to abuse and eliminate rivals. Their acts are scrutinized minutely by lower-status individuals, none of whom would be credible challengers on their own but who in coalition constitute a formidable countervailing force.[3]

In theory, such a society could treat abundant sex as a form of pressure-valve against violent tensions and a reward for peaceable cooperation, much as bonobo societies do.[4] That is a feasible option for bonobo societies because most of the costs of a bonobo pregnancy fall on the mother, who is the one who controls access to the sexual rewards in the first place. She does so because (as in most species, most of the time) males demand as much sex as females are willing to concede them, and the female decision therefore determines how many sexual acts take place.

But the evolutionary niche occupied by early *Homo sapiens* meant that every pregnancy imposed large costs on fathers, grandparents, neighbors, and siblings, on the whole supporting team that each newborn required for its survival. In a world without contraception and delicately dependent on maintaining the incentives for cooperation,

> those costs of sex could not be socialized unless the decisions about engaging in sex were socialized too. The human police state was therefore extended to every aspect of its members' sexual lives. The sanctions it imposed were more severe for women than for men because women, as the more selective sex, were the ones who controlled access to sexual rewards.
>
> It's well known that in police states people spend much more time talking and thinking about transgressions than actually engaging in them. Indeed, talking and thinking about transgressions become a form of intense vicarious pleasure that is compatible with high-pressure group living because it usually has much milder social consequences than those of the transgressions themselves.[5]

These features, in a nutshell, are what likely strike our cousin apes about the links between these three quite un-apelike aspects of our otherwise very apelike human societies. Still, even their professional objectivity might face a sore test when they tried to make sense of the Bill Clinton–Monica Lewinsky story, in which for a couple of years in the mid-1990s the world's most powerful country placed its economic policy, domestic policy, and foreign policy second in importance to determining whether the highest-ranking alpha male in the entire world had engaged in consensual oral sex with someone other than his official partner.[6] Had nobody told *Homo sapiens* that oral sex doesn't lead to pregnancy? Wasn't this a case where the vicarious pleasures of censoriousness had massively larger social consequences than whatever pleasures the participants might have found in the original acts? Was the sexual police state starting to spiral out of control?

Contraception and Consequences

The widespread availability of contraception in the modern world does not mean that sex now has no consequences—far from it. But it does have very different consequences from those it had in the hunter-gatherer communities in which both our social habits and our emotional reactions to each other's behavior first evolved. And it's important to ask whether those habits and reactions are as appropriate now as (in the story helpfully offered by our ape anthropologists) they might once have been.

We are incorrigibly curious about the sexual behavior of others. We also move at lightning speed to judgment, reserving distaste and disapproval for those who are more sexually active than we are (or more sexually preoccupied, or merely differently so), and condescension and scorn for those who are less active.[7] These attitudes may even be self-reinforcing, since our disapproval is heightened by the suspicion that those who are more active than we are feel condescension and scorn for us, and our scorn is likewise heightened by the suspicion that those who are less active regard us with distaste and disapproval. Our reactions sometimes seem to be on a hair trigger, so that behavior that provokes admiration in one context may just as easily provoke disgust in another. All of this makes sense in the context of our evolution. In hunter-gatherer societies the sexual behavior of others had potentially massive consequences for us, not just because it might increase the number of mouths to feed but also because sexual behavior led to the forging and breaking of coalitions on which the very lives of our ancestors depended. It's not surprising that the sex lives of others, and the possibilities opened and closed to us by our own sexual lives, should have become an object of intense reflection and introspection.

Not only do we have intense desires, but we also have desires *about* our own desires and the desires of those we are close to. This is true not just about jealousy. In no other domain is there so much anxiety about how to maintain and stimulate desire. People don't worry about how they can enjoy their food more, or how to savor power. They don't buy pharmaceuticals or concoct ancient herbal remedies to stimulate their desire to watch football, or even to induce their partners to watch football. Our worries about the adequacy of sexual desire are not some dysfunctional by-product of our large brains (it's not because our cortex hasn't got enough to do that we find ourselves contemplating desires about desires). They are the direct result of natural selection. Our sexual partnerships founder without mutual desire, and we are a species for whom life is about partnerships, if it's about anything at all. In hunter-gatherer societies, sexual partnerships were the fulcrum of social cooperation, so of course we care about the desire on which that cooperation hinged.

In a world with contraception, and in which partnerships are built on many more affinities and for many more diverse purposes than in hunter-gatherer society, it seems likely that the intensity of our focus on desires about desires may be misplaced. Infidelity is less serious the less likely it is to lead to pregnancy, and the inadequacy of sexual desire is less serious the more alternative foundations of affinity we have in our social lives. Our inability to treat either of these possibilities less seriously than we used to is likely to be causing us needless distress. Our emotions, which started out as natural selection's way of directing our attention to things that mattered for our fitness, have become the things that matter in themselves, and they still matter even when the fitness landscape in which they used to help us navigate has changed beyond all recognition. As we saw in chapter 3 in relation to the somatic markers hypothesis and the role of emotions in signaling commitment, the relative inflexibility of our emotional responses was not necessarily a disadvantage during prehistory and in some circumstances was positively adaptive. It is likely to mean, though, that our emotional reactions to the circumstances of the modern world are still colored by traces of our prehistoric past.

At the same time, changes in society due to contraception have also brought men and women into proximity in their working environments in new ways. Women have always been at risk of sexual assault from men (domestic servants in particular must have been very often coerced); but the context in which women face sexual coercion has changed significantly. Sexual assault is now a greater risk for women when they are trying to do the same job as men: in particular, they may have fewer other women in a position to act as witnesses on their behalf. Certainly, the conditions of hunter-gatherer societies often kept sexual coercion in check (outside the marital relationship, at least) because women typically forage in groups.[8] Agricultural societies, with their hierarchy and their dependence on slaves and servants, created massive opportunities for sexual coercion, though it was still rare for men and women to work together in conditions of privacy. Now women are much more likely to find themselves alone in the company of men in a way that facilitates both consensual and coercive sexual encounters. Men can use (or threaten to use) economic power or physical force in order to coerce women, and to keep the fact of coercion secret, in ways that would have been difficult in the intimate setting of a hunter-gatherer group. The same police state that overreacts to consensual transgressions of its norms has great difficulty devising an adequate reaction to the continuing prevalence of sexual coercion.[9] Indeed, there are many examples in human societies of sexual coercion being explicitly encouraged as a means of enforcing male norms on women and of punishing women who are inclined to challenge these norms.[10]

Contraception is not the only technological change that has turned our sexual lives inside out. Another truly radical change is photography, and specifically the technology that has made it possible to transform our public as well as private spaces into a panorama of sexual signals entirely detached from the intentions of those who once sent them. Not only can we imagine ourselves to be receiving sexual signals from people we have never met and who care nothing for us, but we can even be distracted by signals where we would normally never expect them. While driving down a busy street, I may receive what seems like an unmistakable sexual invitation from a smiling woman dressed only in her underwear. It is indeed an invitation; but it's an invitation to go shopping. After repeated exposure, I have now rewired my brain to reinterpret the signal, but it's not surprising that some people find that adjustment difficult. One of the reasons men from more conservative societies have often considered modern capitalism to be revoltingly sexually corrupt is that they repeatedly, and understandably, confuse invitations for shopping with invitations for sex.[11] Among the many things they thereby fail to grasp is the extent to which public nudity desexualizes the body. A woman can be represented, or interpreted, as nude and bored, nude and indifferent, nude and oblivious. In the cacophony of the modern street, she is usually all three at once. This doesn't make her images, and the use made of them by advertisers, a sexually trivial matter—but it imposes difficult interpretive demands on our stone-age brains.

Changes in communication technology have had a no less revolutionary impact on our sexual lives. Industrialization radically expanded people's opportunities to meet strangers, compared to their previous lives in agricultural villages (just as agriculture radically changed the opportunities faced by hunter-gatherers). The invention of the telephone added a new dimension to those opportunities, which the Internet and mobile telephony have recently revolutionized once again.[12] Every opportunity to meet a stranger is an opportunity for exchange—of ideas, of economic resources, of sexual favors. It's no wonder our hunter-gatherer emotions find it all puzzling, intriguing, arousing, terrifying. And our responses to the behavior of others—the part we all play in the sexual police state created by our hunter-gatherer ancestors—have now been magnified by those same changes in communication technology and can now resonate around the world.

Lessons from Evolution

Sex need no longer result in pregnancy, and pregnancy need no longer interfere with a woman's capacity to engage in valuable economic activity if she wants to, but many of our instinctual responses to both sex and pregnancy continue to shape our social world. What does that mean for us today? Can we modify these instinctual responses, and should we try?

It's common to claim that whatever we may learn about human evolution cannot determine what we ought to do, because moral prescriptions cannot be logically derived from facts or hypotheses (you cannot derive an ought from an is). While true, that claim is not very interesting, because many scientific facts do indeed give us a basis for action, even if that basis is not one of logical deduction. For instance, the discovery that smoking substantially increases the risk of getting lung cancer gives people a reason not to smoke. It's not that you can logically derive the injunction "Don't smoke" from the factual claim. Instead, the factual claim provides a reason not to smoke for

Advertisement for the healthiness of smoking, 1946. © Apic / Getty Images.

people who already accept a model of reasonable action that puts a good deal of value on staying healthy. Hard to believe though it may seem now, it was once common for cigarettes to be advertised as positively beneficial to health (advertising slogans included "Give your throat a vacation—smoke a fresh cigarette," "Light an Old Gold instead of a throat treatment!" and "More doctors smoke Camels than any other cigarette").[13] Scientific knowledge has made it impossible for such claims to be taken seriously today. As for people who don't place value on staying healthy, it's not clear that there's much to be gained from arguing with them anyway.

The evolutionary history of *Homo sapiens* yields very few conclusions about social behavior that are anywhere near as firmly established as the hypothesis that smoking causes lung cancer. Nevertheless, it has led to an accumulation of evidence that casts into serious question several commonly held beliefs about the relations between men and women that continue to shape how we think about our personal lives and about public policy. These beliefs fall into two main groups, which we can call the model-relationship view and the divided-workplace view. The first consists of beliefs based on the idea that there exists a model form of sexual partnership that can avoid conflict between men and women if we only follow its instructions. The second consists of beliefs based on the idea that men's and women's relationship to the world of work must remain different because of unavoidably different talents, aptitudes, or constraints. Scientific evidence hasn't conclusively proved either idea to be wrong: it has just chipped slowly away at the foundations of their credibility. Let's look at them in turn.

The Model Relationship

Many people believe that nature has made it possible for men and women to have reasonably conflict-free relationships if they only follow certain rules. Many of the world's religions treat marriage as a sacrament conceived and guided by God. Those who think of marriage as a human and not a divine institution often believe that passionate human relationships have been shaped by natural selection in the same way that natural selection has shaped the organs of the human body, to be exquisitely fitted for the purpose of creating and raising new human beings. Such relationships, according to this view, rest on some basic affinities: emotional and intellectual companionship, erotic passion, sexual fidelity, a deep capacity for empathy between partners, and a natural willingness for each partner to take the other's interests as their own. Even if marriage in former ages was an economic contract that gave little space to erotic passion, in the United States and in other prosperous democracies in the twenty-first century it is increasingly seen as both permitting and demanding very high standards of aspiration along many dimensions of the relationship.[14] Single people may take a lot of time to find the right person for a permanent partnership, but once they have found someone, if the partnership is "right," love will ensure that the fundamental interests of the partners are more or less aligned.

In this view, erotic courtship generates sparks that either die down very quickly or ignite the flames of true love: there is no intermediate state. If the fire really does take off, its very uncontrollability, the sense it gives of sweeping all before it, makes it easy to imagine that passion has flamboyantly driven out reason and that true love inhabits a world far removed from the charmless compromises of daily self-interest. We can no more subdue our erotic passion than we can will our heart to stop beating. It's tempting, then, to think of erotic passion as having a design as intricate and as minutely adapted to our survival as the physiological design of the human heart.[15] It's tempting also to think that negotiating and compromising in the pursuit of passion would entail placing our own petty interests above those of our lovers, which would be tantamount to demonstrating a failure of passion itself. If serious conflicts of interest later arise, if there is infidelity, if mutual erotic desire disappears, if the partners are less than fully honest with each other, the relationship is faulty and quite probably was not based on "real" love in the first place. Some (though not all) of those who hold this view also believe that the ability to make a passionate relationship work is a litmus test of honesty and decency: we can therefore judge people's fitness for public office in part by the success of their marriages or civil partnerships.

The biology of human evolution strongly suggests that this model-relationship view is mistaken and that countless couples may be suffering needless unhappiness because they continue to believe in it despite the evidence to the contrary. Nothing in the logic of natural selection is incompatible with sexual conflict. On the contrary, conflict is written into the very heart of the sexual relationship, and it occurs because of, not in spite of, the huge potential for cooperation on which that relationship is founded. Natural selection does not fashion optimal relationships, not even in the limited sense in which it has fashioned optimal physical hearts. In particular, nothing in evolution would have selected for relationships that last a lifetime. If relationships do last a lifetime, it is because the parties can be lucid and constructive about reconciling their conflicting interests. Sexual infidelity is widespread in nature, among females as well as males. That in itself says nothing about how good or bad it may be (murder is also widespread in nature). But the fact that infidelity occurs among individuals who love and want to stay with their partners, as well as among those who do not, implies that it is not incompatible with relationships that last. Couples who wonder why conflict persists between them and suspect it is because they are not "really" in love; single people contemplating whether to commit to marriage or a permanent relationship and trying through introspection to decide whether what they feel is "the real thing"; people whose partners have been unfaithful and who think their relationships have been irremediably tarnished; couples who think that waning erotic desire must be a signal that something invisible is going seriously wrong in the rest of their relationship—all these people may be perplexed or unhappy, and for valid reasons. But they may be making their unhappiness far worse than they need to by comparing themselves to a standard they falsely believe to be natural and possible.

Love may be a many-splendored thing, but its splendors do not include the abolition of conflict, so the persistence of conflict tells us nothing about the nature of love. The fact that your partner is completely unpersuaded by your political views says nothing about how much he loves you: love may make him want to give you his body but not to rent you his brain. If you've convinced her to become a vegetarian for your sake, the fact that your partner continues to crave meat tells you something about the robustness of her appetite but nothing at all about the robustness of her love. If the fact that she continues to be drawn to other sexual partners means you can no longer bear to live with her, leave her if you must, but don't claim it's because you've realized her love for you isn't real. If he's less than honest with you on something you care about, reproach him for his dishonesty, but don't presume he can't really love you: be lucid enough to realize that the greater his love, the greater his incentives to exaggerate or dissemble in its pursuit.

Although jealousy, especially male jealousy, is sometimes regarded as a violent and primitive emotion that is not under voluntary control, it is highly sensitive to social context, and we know that people who in one setting would be "overcome" by jealousy may find themselves capable of mastering their emotions when it is in their interest to do so.[16] As we saw in chapter 4 when discussing the evidence that females had multiple partners during prehistory, various cultural practices can help to soften the disruptive potential of male jealousy. Indeed, male emotional responses to a woman's multiple sexuality are complex, involving not only insecurity and potential aggression but also some element of arousal and attraction (after all, much pornography aimed at men depicts women engaging in sex with other men, and its purpose is to arouse men sexually rather than to provoke them to violence). Different ways of portraying women's potential for multiple sexuality can tip the balance delicately between seeing it as threatening and seeing it as glamorous or charming, as in this commentary from Sam Wasson's book about Audrey Hepburn and the filming of *Breakfast at Tiffany's*. What exactly was it about the famous "little black dress" that Hepburn turned into an unforgettable symbol of chic glamour? "Centuries ago, black dye was affordable only to the very rich. . . . [I]n the Victorian era . . . it was worn almost exclusively by those in mourning. . . . [I]t was a surefire sign of widowhood. To the men passing by, it signified the wearer's knowledge of sex. It meant experience. No wonder the flappers of the 1920s were drawn to it." Wasson observes that to Hubert de Givenchy, designing a dress for Audrey Hepburn to wear in *Breakfast at Tiffany's*, "seeing how this was a dress to be worn by Audrey Hepburn—and not at night, but in the very early morning—it was unusual to say the least. Because it's Audrey—wholesome, wholesome Audrey—there is irony in her endorsement of a color heavy with unchaste connotations. . . . [T]he contrast is sophisticated. Black on Audrey Hepburn gives her an air of cunning . . . that's the essence of glamour."[17] In short, in the right cultural context and with a sufficiently deft touch, male jealousy can be channeled into something a lot more playful and a lot less threatening.

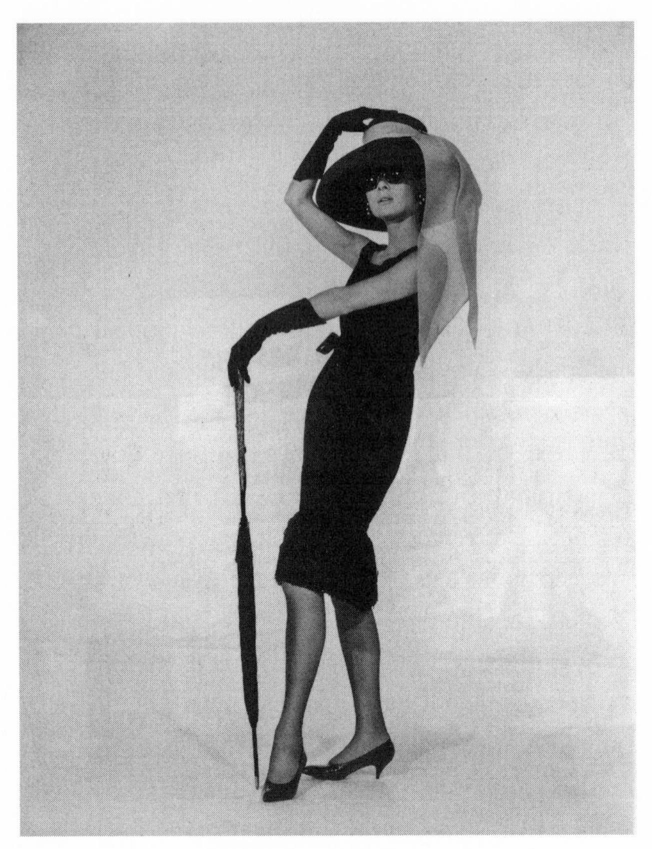

Audrey Hepburn and the "little black dress" designed by Hubert de Givenchy for her to wear in the film *Breakfast at Tiffany's*, 1961. © Hulton Archive / Getty Images.

The model-relationship view also has a powerful impact on politics in many countries, making candidates for public office feel obliged to project images of themselves as good spouses and parents. In some countries (notably the United States), the breakdown of a marriage can often damage, and sometimes end, a person's political career. I'm not aware of any scientific study that has shown that people whose relationships last, and who avoid infidelity, are any better at serving in political office than people who don't, and it would be hard to imagine how such a study would be conducted. Since even the pretense of scientific evidence is almost never offered in such cases, it may be that those who join the hue and cry after politicians whose relationships break down are not really persuaded by evidence at all. They may merely be indulging the pleasure of censoriousness that we've already seen to be a highly evolved part of our nature, as well as using against political opponents the notoriously hair-trigger character of our disgust at other people's sexuality.

In any case, if there were such evidence, it would almost certainly indicate that the case for resigning from politics when your marriage breaks down should apply also to business, journalism, and indeed most other professions. The toll such resignations would take on productive organizations is terrifying to contemplate. In the absence of evidence, it's probably best to consider the periodic outbursts of public indignation at the relationship misfortunes of others as a kind of blood sport. Our delight at interfering in the consensual relationships of others would be amusing if it did not divert our attention from the much more serious (and much more difficult) problem of sexual coercion.

In short, the model-relationship view may represent some people's ideal, but an ideal has to be attainable if it is not to paralyze with self-reproach the people who fail to live up to it and make it harder to be lucid about living with relationships that do not conform to the model.[18] The biology of human evolution indicates clearly to us that the persistence of conflict among couples need not mean the end of love, that deception or infidelity need not mean the end of a relationship, and even that the end of a relationship need not mean the end of someone's world. To say this is much more than to repeat the truism that no relationship is perfect: it is to reject a specific biological model of relationships as being as well adapted as organs of the human body. A cardiologist trying to repair your physical heart needs first to understand how the model heart works. The repair of emotional heartbreak needs a different approach, one less wedded to an ideal model of common interest and more open to the lucid expression of what the partners want and need from each other, even and especially when these wants and needs are in conflict.

The Divided Workplace

The persistence of gender disparities in salaries in many occupations, as well as the persistent underrepresentation of women in powerful positions in business and public life, leads certain people to conclude that there are things about modern working life and about being a fulfilled woman that do not quite mix. We saw in chapter 4 that in prehistory, there was a deeply ingrained division of labor between men and women that resulted in entrenched economic inequalities. The economic basis for that division of labor has now completely disappeared, but the inequalities, though much diminished, are still with us. We saw in chapters 5 to 7 that there is no evidence that these disparities have anything to do with differences in talent between men and women. Instead, there is substantial evidence suggesting they are due to a combination of two factors:

1. Differences in preferences for which women pay a high (and probably unreasonably high) price. Taking career breaks and working fewer hours than many men appear to incur particularly large disadvantages. These choices have an adverse effect on women's advancement not just in the child-rearing years but for decades afterward. It makes a difference also in that men are rewarded, as women are not, for more aggressive bargaining strategies in the professional environment.
2. Subtle differences in networking strategies between men and women that can operate to make the talents of women less conspicuous to potential colleagues and employers than those of equivalently talented men.

These two factors are not alternative explanations but complementary ones, in that the second factor explains how the first can be so powerful and persistent. It is the differences in networks that explain why entrepreneurs do not easily move to take advantage of the opportunities offered by talented but underrewarded women in the modern workforce. The reason why women pay such a high price for their choices appears to have something to do with a signaling trap, in which it is difficult for women to signal their talents and their motivation consistently with organizing their lives to accommodate the many goals they can and should want those lives to accomplish. Men often pursue a less diverse set of goals, partly because of social conventions that encourage them to do so, and a by-product of this more single-minded strategy is that their signals are easier for employers and colleagues to interpret.[19]

Can greater understanding of the causes lead to a less divided workplace, and should we want it to? Obliging or pressuring women to act against their preferences is hardly to respect their dignity or their autonomy. Two things suggest that a better understanding of the causes of the divided workplace may lead naturally to some softening of the divisions, without any compulsion or bullying involved. First, some of the apparent differences in preferences between men and women are likely to reflect differences in their constraints that may erode over time as the true costs of those constraints become clear. For instance, the professional single-mindedness that appears to be preferred by many men may be due partly to their being more able than women to find flexible partners who accommodate them. Some professional women might indeed be as single-minded as their male colleagues if only they could find partners like theirs. As the title of Terri Apter's book neatly expresses it, working women don't have wives.[20] If the ability to pursue a career single-mindedly without sacrificing a home life becomes something that any individual needs to negotiate openly with a partner, rather than presuming it as so many men have done in the past, gender differences in these choices may persist, but there may be less stark differences than we see now. And those differences that do persist may be easier for both parties to live with if they are chosen rather than imposed.

Second, merely understanding better the causes of the divided workplace may lead to more imaginative initiatives on the part of employers and colleagues. For instance, understanding the fact that individuals who have taken career breaks are underrewarded compared to similarly talented men may lead more employers to seek out such women systematically. They have a particular incentive to do so if they realize that the lower rewards such women receive have everything to do with their lack of conspicuousness once they drop out of male networks and rather little to do with any lack of talent or commitment revealed by their career breaks per se. They may also come to see the virtue of recruiting outside the networks of people who have spent their entire lives working for the same kind of firm without a break, and who may therefore be blind to the complacent assumptions that industry insiders routinely take for granted. True, corporate cultures are intrinsically conservative, and corporate recruiters are not very good at looking systematically for recruits who differ in important ways from themselves. That's why competition—and especially openness to entry from firms outside the industry—is important in stimulating innovation in all dimensions, in recruiting procedures no less than in products. But in a signaling trap, expectations are important too. If people who work in business routinely expect to see all-male boardrooms, that makes it easier for them to justify making little effort to recruit outside the circles they know. If they start to understand—from the way shareholders talk to them at shareholders' meetings, from the reactions of clients and customers, from what their colleagues in other firms report, from what they read in the press—that there's something a little odd about choosing your senior executives from only half of the available talent pool, then even the most complacent recruiters might begin to feel uncomfortable with the status quo.

Coupled with the increase in the educational attainments of women compared to men in recent years (described in chapter 8), innovative recruiting initiatives may mean that over the next two or three decades, the differences in economic power between men and women will largely disappear.[21] Still, among the talent pool of individuals in their forties and fifties from whom today's corporate boardrooms are recruited, women are already slightly more educated, on average, than men.[22] So if the startling disparity in their representation in senior positions (women make up only 2.4 percent of Fortune 500 CEOs, for instance) can persist in these conditions, there's no guarantee that it cannot hold out against the even larger disparity in educational attainment that will appear over the next twenty to thirty years. The evidence reported in chapter 8 suggests also that there might even be some tendency to appoint women to nonexecutive positions to avoid addressing the imbalance in appointments to executive jobs, where the real power in organizations is exercised. So even an advance in women's representation in visible positions, such as nonexecutive directorships, might not do much to shift the balance in the executive positions.

Can or should public policy realistically do anything about this? I suggest in chapter 8 that a fundamental challenge for anyone in a modern economy is to attract and keep the scarce attention of potential employers, clients, and colleagues in a world of information overload. If women are to compete equally with men, they need to overcome what appears to be an attention deficit, to make their talents as conspicuous as those of equivalently talented men. I've already suggested that many of the most appropriate actions to redress this deficit should be undertaken through individual initiative rather than public policy. It's certainly appropriate to be wary of wielding clumsy policy instruments to deal with a very subtle and elusive problem. Still, two kinds of public policy have often been suggested to complement the kinds of antidiscrimination provisions that are already part of the legal framework in many countries, and it's worth examining both of these in greater detail.

The first type of policy aims to address imbalances in the attention of recruiters. Requiring simple parity in employment is unworkable: it is not only unrealistic to require that firms employ equal numbers of men and women, but even if it could be achieved, it would be at a large cost in flexibility and adaptability, virtues that firms in the twenty-first century need more than ever. But it's much less unrealistic, and much less intrusive, to require firms to choose the men and women they employ from short lists that reflect a reasonable balance of the sexes. The reasonable balance would vary according to the occupation: when only 1.3 percent of airline pilots are women, it won't be feasible to insist there be 50 percent of women on every short list for recruitment of pilots, but it might be feasible to insist that every short list contain at least some women (unless it can be shown that in a wholly open recruitment process, none have applied). Balanced short lists also have a major advantage over gender quotas: nobody need feel they got their job only because of the quota. Even if it's true of some women that they got the job because the quota ensured they made it to the short list (which will necessarily be the case, otherwise the policy would be ineffective), it will still be true that they beat the other candidates on the short list in a fair fight.

Of course, such a policy would be completely ineffective if firms systematically treated it as window dressing, adding women to their short lists but not appointing them to jobs in any greater proportions than before. But the evidence we've discussed up to now does not suggest that firms are systematically trying to exclude women: their actions are having that effect largely through a massive failure of imagination, a failure to spot the pool of talent that exists just outside the field of vision of their insular, self-perpetuating networks. If that's so, requiring more women on short lists would help to shift more of that talent inside the corporate field of vision, and the evidence suggests that this might substantially alter recruitment outcomes. But it's important to push the argument to its logical conclusion. If such a policy worked, it would not need to be compulsory: it would be in companies' own interest and could therefore equally well be implemented through voluntary codes of practice and through certification procedures that signaled to the outside world that firms take the gender balance of their recruitment procedures seriously. A quality label that firms could attain only by systematically gender-balanced short-listing might well be a sufficiently attractive signal in the labor market that firms would want to make such commitments without any legal compulsion.

A second type of policy aims to undermine the signaling trap by ensuring that career breaks taken for the purpose of raising children are not interpreted as a lack of motivation or talent for employment. Compulsory paternity leave that precisely matches the availability of maternity leave is one such measure.[23] There is a rationale for paternity leave that has nothing to do with its effect on the gender balance in employment: namely, the idea that parents have a responsibility to the rest of society for the way they bring up their children, because of the large externalities, positive or negative, that those children will create for their fellow citizens as they grow up. It's therefore quite reasonable for society to insist that bringing up children, which is an area where private decisions have wide social repercussions, is a paternal responsibility as well as a maternal one, and the provision of paternal leave is one way to make that point clear. Although public policy cannot ensure that fathers use their paternity leave wisely or productively, it can establish a convention that parenthood involves career interruptions regardless of the talent and motivation of the person concerned. Fathers who believe that their paternal obligations can be fulfilled without any diversion whatever of time from their professional to their domestic responsibilities have misunderstood the nature of what their fellow citizens require from them.

Although the justification of such a policy rests on the fact that paternal responsibilities have repercussions for the rest of society, there may be an additional benefit for gender representation, one that arises from the impact on the signaling trap. Once firms wake up to the fact that most of their potential CEOs will be people who have taken parental career breaks, they're likely to stop considering the decision to do so as revealing qualities that are undesirable in a CEO. It won't stop some parents from taking longer career breaks than others, and it probably won't equalize the gender distribution of those career breaks any time soon. It is not easy to find evidence about the likely impact of such policies, because in the few countries that have tried anything like them, the take-up of paternity leave has been largely voluntary. This makes it hard to compare the behavior of those who take it up with those who do not, since many other factors (including the motivation of the individuals concerned) are likely to be varying at the same time. However, one study investigating this question was conducted in Norway (where paternal leave was made highly advantageous in a reform in 1993, so that take-up of leave by eligible fathers increased from 4 percent before the reform to over 60 percent within a couple of years). Interestingly, this study suggests that compulsory paternity leave may increase the tendency of women to take time off work to rear children, perhaps because it becomes more pleasant to do so once the task is shared.[24] The idea that paternal career breaks might simply substitute for maternal breaks is not borne out by this experiment. If anything, the message is that both men and women can put a higher weight on their domestic responsibilities when they can do so at a lower cost to their careers. Nevertheless, such a measure may help, through changing the signaling equilibrium in employment, to ensure that those who take career breaks pay a price that more nearly reflects the balance of costs and benefits to society as a whole.

As both of these examples probably suggest, the role of formal legislation in redressing the gender imbalances that I've highlighted is only a part, and probably a very minor part, of the solution. Far more important will be changes in norms and conventions among men and women alike, as we come to realize that methods of organizing work that penalize women for their career choices are not just inequitable—they're also profoundly stupid. The evidence we've looked at in the last few chapters suggests it is likely (not certain, but likely) that there are talented women everywhere who are being paid less and given fewer opportunities than equally talented men. If so, it's time for everyone, women and men together, to wake up to what an opportunity these women represent for their potential employers and colleagues.

Many working practices, such as spending long hours in the office doing tasks that could just as efficiently be performed at home, are a form of wasteful display. They are a peacock's tail of individually rational but collectively pointless effort that happens to penalize women, on average, more than men in terms of professional advancement but penalizes men more than women, on average, in terms of their domestic and family fulfillment. In fact, these costs to both men and women are probably a more important reason for taking action than the remaining inequalities in themselves. What really matters is not that some already highly paid women are paid less than their male counterparts but rather that both men and women are failing to find the fulfillment that a modern productive economy ought to make possible for them.

Escaping the logic of the peacock's tail will not be easy: a peahen that chose to ignore peacocks with large tails would not increase her fitness by doing so. But that's because she would have no other way of discriminating the fitter males from the less fit. If she could find such a better basis for diagnosing male fitness, ignoring those large tails would be a very smart move indeed. If *Homo sapiens* is to wring any collective benefit from the fact that we have larger brains than peacocks, it must surely include the possibility of devoting our ingenuity to escaping such signaling traps by devising less wasteful ways to reveal our talents and motivations to each other. Technology is transforming the physical constraints of our working lives; it does not seem unrealistic to hope that it can reorganize some of the social constraints as well.

The revolution in contraceptive technology, and the revolution in society that has accompanied it, have made it possible for the first time in the history of our species to separate the sexual collaboration between men and women from their more general economic relations. That will not mean, and nor should it, that sex is banished from the workplace. But sexual relations and economic relations can achieve more when each is liberated from the shadow cast by the other. Just as sex freed from economic dependence is usually better sex, economics freed from dependence on sex is likely to be better economics too.

NOTES

Chapter One: Introduction

1. Posner (1994, 111) writes: "The ends [that sex serves] fall into three groups, which I shall call procreative, hedonistic, and sociable. The first is obvious. The second has two cells. One is relief from the urgency of sexual desire; the analogy is to scratching an itch, or to drinking water when one is thirsty. The other is ars erotica, the deliberate cultivation of the faculty of sexual pleasure; the analogy is to cultivating a taste for fine music or fine wine. The third group of sexual ends, the sociable, is the least obvious. It refers to the use of sex to construct or reinforce relationships with other people, such as spouses or friends."

2. See the many studies in Sommer and Vasey 2006, which investigate behavior among bison, cats, dolphins, flamingos, and geese as well as many primates. Not all researchers agree about the reasons for homosexual behavior in nonhuman species, and a single explanation may not apply to all species. There are also interesting unresolved questions about whether humans show a particular propensity for exclusive homosexuality as opposed to bisexuality.

3. Many examples, including several cited in this paragraph, are described and discussed in Arnqvist and Rowe 2005 (e.g., 1–13). The various chapters in Muller and Wrangham 2009 offer a comprehensive summary of what is currently known about sexual coercion, including hypotheses about its evolutionary origins. Judson (2002, especially 9–20) offers an entertaining account of sexual conflict in various animal species.

4. See Barash and Lipton 2001; Birkhead 2000.

5. On lions, see Packer and Pusey 1983. The evidence for many primate species is summarized in van Schaik and Janson 2000, chapter 1, especially table 2.1 and pp. 40–41.

6. Arnqvist and Rowe 2005, 50–52.

7. See Whitchurch, Wilson, and Gilbert 2010. Kim Sterelny has suggested to me that sexual relations are unusual in this respect: "It does not seem to be true of other human co-operative alliances, where there is a lot to be gained and lost. We do not find

potential co-authors more appealing, if we are not certain what they think of us. But writing a book together is a big and risky investment. So what makes uncertainty a bargaining lever in sex but only sex?" (personal communication, July 2011). On reflection I am not convinced that sexual relations are unusual in this respect. In labor markets, higher wages or bonuses are often paid to employees who can credibly threaten to leave; the threat of leaving itself conveys that the employee is more highly valued on the open market and therefore (perhaps) of greater value to the employer. Similarly, lovers whose affections may stray elsewhere effectively communicate that they are considered more attractive by others and may therefore seem a more valuable catch. In a different domain, one of the hypotheses of the attachment theory of parenting is that children need a "secure base" (Bowlby 1988) from which to explore the world. In this view, the more unpredictable the behavior of their caregivers, the more emotional energy the children need to invest in claiming their attention and the less they have available to explore their environment. On the application of attachment theory to adult relationships, see Crowell and Waters 2005 and the various contributions to Mikulincer and Goodman 2006.

8. Not all animals that reproduce sexually have two sexes, and in those that do, it is not always true that individuals are determinately of one sex or the other (some fish, for example, change sex according to the distribution of mating opportunities). These points, and the implication that sexual strategies are far more varied across nature than even Darwin realized, are well made in Roughgarden 2004 and 2009. Here I discuss species in which there are indeed two distinctive sexes and all or most individuals are determinately of one sex or the other.

9. A classic article by Trivers (1972) emphasized that it is the asymmetry of parental investment (including the investment in gestation) that creates these divergent incentives for males and females rather than the asymmetry of gamete size per se.

10. It can sometimes be illuminating to consider relations between the sexes as if they were a market in which males demand and females supply sexual services, a perspective developed, for example, in Baumeister and Vohs 2004. However, this view is a simplification for modeling purposes, and it should not be taken to imply that females do not value sex or the quality of the sexual encounters they engage in.

11. Bowles and Choi 2007.

12. Barash and Lipton 2001 has an extensive treatment of this evidence (e.g., p. 12). Birkhead 2000 focuses particularly on female infidelity and sperm competition (195–231). See Ledford 2008 for the surprise discovery of extra-pair copulations in a species of vole previously considered monogamous. Knight 2002 briefly and accessibly surveys studies of multiple sexuality among females of a wide range of species. Judson 2002, chapter 1, is another excellent source.

13. Baumeister (2010, 221–29) discusses at length the evidence that women on average desire sex substantially less often than men, based on how couples separately report their satisfaction with the frequency of sex within the relationship. Without distinguishing between the quality of sexual encounters, such evidence may be misleading. It's likely, for instance, that women may report sex with their current partners to be frequent enough given the quality of the experience, while still feeling that they might like sex more often if the experience were better—that is, if their partners were more generous and attentive. Men's preference for more sex than they currently get may reflect simply their greater preference than women for mediocre sex over no sex at all. Meston and Buss (2009) emphasize the wide range of motivations cited by women for desiring sex and the many kinds of cues that can trigger such a desire, suggesting that it may not be very informative to compare men and women with respect to an unconditional desire for sex, given how much the desire tends to be conditioned by circumstances.

14. Strictly speaking, all potential fathers might make some contribution to raising offspring, even if they are uncertain about their paternity without that uncertainty implying confusion (e.g., if several potential fathers have subjective paternity probabilities that add up to one). It is in the mother's interest to make each of them believe his probability of paternity is higher than it really is.

Chapter Two: Sex and Salesmanship

1. Centorrino et al. 2011.

2. On the signaling role of dance, see Brown et al. 2005.

3. Manfred Milinski (2003) has shown that scent plays an important role in facilitating mating between individuals whose immune systems have an intermediate degree of difference from each other.

4. Arnqvist and Rowe 2005, 74–77.

5. There is a substantial literature showing that perceived physical attractiveness is positively correlated with labor-market rewards (Hamermesh and Biddle 1994; Mobius and Rosenblat 2006) and electoral success (Berggren, Jordahl, and Poutvaara 2006). This finding is reviewed and its implications discussed in Hamermesh 2011 and Hakim 2011. The literature on signaling to friends and colleagues is vast: Miller 2009 provides an enjoyable, nontechnical overview. Bénabou and Tirole 2006 and Seabright 2009 develop models of the use of prosocial behavior for signaling motives. Ariely (2008, chapter 13) gives an amusing experimental demonstration of how ordering behavior in restaurants may be influenced by signaling motives, at a real cost in the perceived quality of the items ordered.

6. See Robin Hanson's blog, *Overcoming Bias*, September 15, 2010 (http://overcomingbias.com/2010/09/a-med-datum.html) for an excellent (anonymous) contribution from an ophthalmologist's assistant who describes the ostentatious alcohol swabbing of equipment in front of patients, while eye droppers are reused in ways that pose far more infection risk than would exist from unswabbed equipment.

7. There is evidence from both humans and other primate species that individuals perceived as sexually attractive are also treated with greater attention and generosity in nonsexual interactions, including those involving members of their own sex. In addition to the literature cited in n. 5 above, see Sapolsky 2005. Wallner and Dittami (1997) show that female Barbary macaques with larger than average anogenital swellings are not only more attractive to males but are also preferentially groomed by other females. Causality is not easy to establish here: do partners in nonsexual interactions tend to favor individuals that seem likely to enjoy greater sexual success, or do individuals attract more sexual partners because they appear to enjoy more social success? Both alternatives seem plausible.

8. Goffman 1963; Yoshino 2006. As Yoshino acknowledges, even members of majority cultures do something similar to covering, because their identities are multiple and they are continually balancing the requirements of one context against those of another.

9. See Fox 1984.

10. Tungate 2007, 15.

11. The advertising campaign used the supermodel in a number of different contexts, one of which was to have her (apparently) crashing the car deliberately to show off its airbags and other safety features. Here the fact that the advertisers were prepared to take risks with such a famous model would itself have signaled their confidence in the reliability of the car's safety systems. See YouTube video, "Citroen Xsara Advert: Crash Test with Claudia Schiffer," http://youtube.com/watch?v=G71qlOk-qoY, accessed May 18, 2011.

12. See Doniger 2005, especially chapter 1, which discusses Cary Grant along with many other cases.

13. Miller 2009, 72.

14. Veblen 1915, 43.

15. Veyne 1996.

16. Foreman 2009.

17. Sexton and Sexton 2011.

18. Foreman 2009.

19. Calculation based on a return fare of €13,400 (US$16,400) from Paris to Tokyo with Air France, quoted on June 30, 2010, with a total flying time of just under twenty-four hours.

20. See Searcy and Novicki 2008 for a survey and discussion of the problem of honest signaling in song sparrows.

21. Reid et al. 2005. This kind of research is more difficult than it seems, and the difficulty is not simply that of crawling around in the undergrowth trying not to frighten the birds. Strictly speaking, the fact that birds with larger repertoires have more offspring could be due solely to the fact that more fertile females choose birds with larger repertoires. The researchers present other evidence to suggest that the males' contribution is important, notably that males with larger repertoires also live longer on average, suggesting that they have better underlying health.

22. MacDonald et al. 2006.

23. I hope readers will feel that this claim does not need to be corroborated with a scientific study.

Chapter Three: Seduction and the Emotions

1. See Damasio 1996. This is not to say that specific regions of the brain embody specific emotions: the embodiment appears to be functional rather than structural (see Lindquist et al. 2011).

2. This work is surveyed in the papers collected in Ellison and Gray 2009.

3. Damasio 1994.

4. Frank 1988. Fessler and Haley (2003) survey and discuss the role of emotions in human cooperation. I discuss the evidence for the emotions as a positive force for human prosociality in Seabright 2010, chapters 4 and 5, laying particular emphasis on their role in the reduction of violence in modern human societies. Bowles and Gintis (2011) discuss in great detail the evolution of human prosocial behavior.

5. Stendhal 1962, 37 (my translation).

6. Gambetta 2009, 258–59.

7. See Miller 2000, 238–41.

8. Kim Sterelny has pointed out to me that it is an established finding in psychology that "variable reinforcement schedules lead to highly persistent behaviour, so the variability of female experience might equally be taken as an adaptation to ensure continuing sexual interest" (personal communication, July 2011). This is possible, but because the male's continuing interest is typically more important to the female than vice versa, this logic would suggest that we should expect the male orgasm to be the less reliable of the two.

9. I consider evidence that this was so in chapter 4. The argument also evidently requires that no individual sexual encounter have too high a probability of leading to pregnancy.

10. See Caldwell and Young 2006 for a survey of the effects of both vasopressin and oxytocin, Kosfeld et al. 2005 for the social impact of oxytocin on trust in human subjects, and Walum et al. 2008 for evidence that genetic variation in vasopressin receptors correlates with strength of pair-bonding behavior in humans. The physiology of orgasm is discussed in great detail in Komisaruk et al. 2006. Lloyd (2005, especially 107–48) argues in favor of Donald Symons's theory of female orgasm as a pure by-product of male orgasm and claims that rival theories indicate important biases in the largely male community of scientists studying the question. As I suggest here, a by-product view of the origins of female orgasm might be compatible with a view that natural selection might co-opt such a mechanism for certain adaptive purposes, such as screening.

11. See Buss 2003, 223–34. In particular, female orgasm may have had the advantage of also responding selectively to lovers with "good" genes, since orgasm is known to increase the rate of sperm uptake into the uterus.

12. Cunningham, Barbee, and Pike 1990.

13. See Owren and Bacharowski 2001, 156.

14. We report our findings in Centorrino et al. 2011.

15. To avoid spurious correlation, in case individuals who have decided to trust others try to justify their acts by rating smiles as genuine, we compare an individual's trusting behavior with the ratings of smiles by other participants.

16. The effect on trust of smiles in still photographs was first shown by Scharlemann et al. (2001).

17. This concept was introduced into biology by Zahavi (1975) and formalized by Grafen (1990). Similar theories were being developed in parallel by the economists Michael Spence (1974) and James Mirrlees (1997).

18. In the screening model of Michael Spence (1974), workers may be of high or low productivity, but this value is not observable by employers. Employers can, however, observe how much education the employees choose to acquire. If high-productivity employees have lower costs of acquiring education (because they can succeed in class with less work), they will choose higher levels of education than low-productivity employees even if education brings neither them nor their employers any benefit at all (and is a pure handicap). The value of education to high-productivity workers is that it signals not just that those workers can bear the handicap but also that they will have higher productivity when hired by employers.

19. Weatherhead and Robinson 1979.

20. Gibson and Höglund 1992. Galef (2008) summarizes studies of mate-choice copying in Japanese quail: not only is this finding confirmed for females, but the opposite is found for males, which is consistent with the greater scarcity of the female bird's gametes than those of the male. However, the evidence on whether women find

attached men more sexually attractive is mixed: for instance, Uller and Johansson (2003) find no such effect.

21. A variant of this strategy is used (or so is my unscientific impression) by younger women in the company of older and ostentatiously wealthier men: by spending a lot of time telephoning or texting, they can signal, "I'm not really with this guy." This behavior is consistent with the dual strategy discussed by Thornhill and Gangestad (2008), who argue that human females have evolved two distinct sexualities: an "extended sexuality," when conception is impossible, whose purpose is to elicit material benefits from males, and an "estrous sexuality," whose purpose is to obtain "good genes" when conception is a possibility. There may be visibly different tendencies in sexual behavior under these different conditions without the tendencies being entirely distinct: behavior that influences the possibility of better genes would also have a (probabilistic) impact on the behavior of contributing males, and natural selection might have made human female psychology very sensitive to the mingling of these two currents.

22. Darwin 1981.

23. Desmond and Moore 2009.

24. Carroll 2010, 122–27.

25. Coyne 2009, 92.

26. Ridley 2004, 186.

27. Gee, Howlett, and Campbell 2009, 15.

28. Darwin 1856.

29. Shakespeare, sonnet 116. Boyd 2012 offers a powerful analysis of Shakespeare's sonnets that uses the insights of signaling theory, and of evolutionary theory more generally, to convincing effect. This sonnet is one of those addressed to the Fair Youth, so the context is not that of the standard Petrarchan sonnet addressed by a man to a woman, but that does not make signaling any less central to its concerns.

30. See Cooper 2008.

31. Trivers 2000, 115: "True and false information is simultaneously stored in an organism with a bias towards the true information's being stored in the unconscious mind, the false in the conscious. And, it is argued, this way of organizing knowledge is oriented towards an outside observer, who sees first the conscious mind and its productions and only later spots true information hidden in the other's unconscious." He also suggests a different theory of self-deception, suggesting that it arises through the manipulation of components of our beliefs by others, such as parents. Trivers 2011 develops his views on self-deception at book length. The experimental evidence that tends to support Trivers's first theory is reported in Valdesolo and DeSteno 2008.

32. There is an abundant literature in psychology on motivated reasoning, where the desire to reach certain conclusions influences cognitive processes to an extent that

is constrained by the need to provide plausible rationalizations for the motivated beliefs (see Kunda 1990). Haidt 2007 suggests that this plays a fundamental role in moral reasoning, since much explicit moral argument seeks to confirm the soundness of pre-existing moral intuitions. Whether empirical or ethical, such forms of reasoning pose an obvious question: why did natural selection make us that way, instead of giving us reasoning faculties much more focused on seeking out truth? There is no consensus about this among researchers. Trivers's theory is one suggestion whose applicability is limited to truths about facts it would be strategically valuable to conceal from others. Haidt believes that motivated reasoning in morality is a by-product of recent evolution: "Language and the ability to engage in conscious moral reasoning came much later, perhaps only in the past 100 thousand years, so it is implausible that the neural mechanisms that control human judgment and behavior were suddenly rewired to hand control of the organism over to this new deliberative faculty" (2007, 998).

33. It now seems clear that the pain of rejection by a lover is a real, physical pain that activates brain regions similar to those activated by painful physical stimuli; see Kross et al. 2011.

34. For a wonderful tour of the confusions and paradoxes of sexual signaling, see the three books by Wendy Doniger (1999, 2000, 2005): the second is most explicitly about sex, while the first is about gender and the third about love.

Chapter Four: Social Primates

1. The original Latin phrase is "Bellum omnium contra omnes." It appears in Hobbes 2008, chapter 1.

2. Darwin, 1979, chapter 14.

3. Darwin 1981, 162–63.

4. The literature on primate societies is vast; an excellent place to start is de Waal 2001.

5. Marmot 2004. Recent work by Armin Falk and coauthors (2011) has shown that in experimental settings, receiving pay that is perceived as unfairly low increases heart-rate variability, which is known to predict cardiovascular disease in the long run. They also show in data from firms that there is a strong association between cardiovascular health and perceived fairness of pay.

6. Muller and Wrangham 2004.

7. de Waal 1982, 138. Kim Sterelny has suggested to me that the relative instability of the coalitions in chimp societies, compared to the relatively norm-driven and institutionally structured social worlds of humans, can be compared to the contrast between market transactions without fixed prices, in which everything is continually negotiated, and fixed-price shopping (personal communication, July 2011).

8. Gesquière et al. 2011.

9. Harcourt et al. 1981. Most of the relevant research has been done in primates, but there's no reason that the relationship should be confined to primates. The very large testes of whales and dolphins (those of the right whale weigh over a ton) suggest likely polygyny: see Ridley 1993, 220. This relationship holds between species rather than within them: there's no evidence that men with larger testicles are able to have sex with larger numbers of women (despite the claims of the many websites devoted to techniques for testicle enlargement).

10. See Simmons 2001 for an account of sperm competition in insects.

11. Wrangham and Peterson, 1996, 225–27.

12. Stanford 1999, 42.

13. See Leduc 1992 for a discussion of marriage as gift exchange in classical Greece. Leduc makes the very interesting point that it was in the more socially conservative city-states, such as Sparta and Gortyn, that women could be citizens and own property in their own right; indeed, citizenship derived from membership in a community of landowners that was closed to outsiders. In Athens, which was socially more innovative and inclusive with respect to male foreigners, the locus of citizenship remained the (male-headed) household, and women passed from father's to husband's household, treated not so much like chattels as like children. As she memorably expresses it, "Women were the chief victims of the invention of democracy" (239).

14. Baron-Cohen 2003.

15. Croson and Gneezy 2008.

16. There is much controversy on this topic. That measured IQ scores show significant racial differences on average is not in serious dispute. What is controversial is the extent to which these differences are attributable to genetic variation, both because of selection effects and because there is strong evidence for large environmental variation of the relevant kind. Rushton and Jensen (2005) have argued that the proportion attributable to genetic variation is large; Fryer and Levitt (2006) use tests on very young children (where observable racial differences in scores are very small) to argue that this is unlikely. Hunt (2011, chapter 11) has a good, balanced overview that emphasizes how little we still know about this difficult question, and I discuss the deficiencies of IQ as a measure of intelligence in chapter 5. Here I want merely to make the point that the logic of sexual selection would lead us to expect very small differences in intelligence between human populations but potentially large differences between men and women. It is therefore both surprising and fascinating how hard the latter differences are to find.

17. Darwin 1998.

18. See Wilder, Mobasher, and Hammer 2004; Seabright 2010, 6.

19. Seabright 2010, 5–6. Cochrane and Harpending (2009) discuss ways in which relatively recent human evolution has been rapid, though these are consistent with the argument I advance here.

20. The Flynn effect (see chapter 5) shows how large are the likely environmental effects on test scores, and how small by comparison are the differences in scores between populations.

21. See Baron-Cohen 2003, for example.

22. Sterelny (2012) discusses these changes at length, emphasizing how tiny cultural modifications accumulated over a long time can add up to major qualitative change. Some innovations had an important effect on the balance of power between men, such as the development of projectile weapons, which made physical contests less dependent on simple strength. The gender balance of power may have been affected by other innovations, such as the making of fabric, which from the beginning appears to have been an essentially female occupation (see Barber 1994). Adovasio, Soffer, and Page (2007) have argued that this and other technologies manufactured by women (including ropes and nets) played a central role in human social evolution.

23. Ryan and Jetha 2010, 12.

24. See Mead 1973; Freeman 1983. In fairness to Mead, it should be said that social anthropology in the early twentieth century combined fascination with sexual practices of non-Western cultures with a free-and-easy approach to evidentiary rigor. A flavor of the latter can be found in Ernest Crawley's *Studies of Savages and Sex* (1929), with its disquisitions on such topics as "the sexual impulse of the savage," in which Crawley makes statements such as "Even the negress is by no means very amorous" (9), on the authority of Havelock Ellis's reference to "a French army surgeon familiar with the black races in various French colonies." Mead's evidence gathering was comparatively thorough.

25. Ryan and Jetha 2010, 217.

26. Hrdy 2009, 155.

27. See Thornhill and Gangestad 2008.

28. Cornwallis et al. 2010.

29. Dunbar 1992. The relationship, though present in the data, is not statistically very robust: the hypothesis that living in larger groups requires greater neural processing capacity, and might therefore lead to selection for larger brains, rests on a range of evidence, much of which is summarized in Sterelny 2003.

30. See Kaplan et al. 2000; Hooper 2011, especially figure 3.7.

31. See Hawkes et al. 2000. There is controversy about how central grandmothers were to human social evolution: see Sterelny (2012, especially chapter 4.3) for a view that makes them part of the story but a less crucial part than claimed by Hawkes et al.

32. Wrangham et al. 1999; Wrangham 2009.

33. See Hawkes et al. 1991; Kaplan et al. 2001.

34. See Hawkes 2004, for example. This point does not invalidate the example, since males still exert substantially more control over females than in other ape species, even if the control remains incomplete and variable across cultures and circumstances.

35. Hrdy 2009, 151.

36. Copeland et al. 2011; Alvarez 2004.

37. Hawkes 1991.

38. Boehm 1999.

39. Bowles 2009. Seabright 2010, especially chapters 3–5, discusses the evidence for violence in forager societies and explanations for the much lower modern levels of violence. This is also the theme of Pinker 2011, which documents the downward modern trend in violence in compelling detail.

40. Steckel and Wallis 2009, table 2.

41. Seabright 2010, 265 and n. 2.

42. See Stearns 2000, chapter 1.

43. Diamond 1987.

Part Two: Today

1. World Health Organization 2011, file DTH6 2004.xls.

2. World Health Organization 2011, file vid.680.xls.

3. See Hill, Hurtado, and Walker (2007, figure 1), on the Hiwi of Venezuela. Gurven, Kaplan, and Supa (2007, table 5) report similar evidence for the Tsimane of Bolivia, as do Hill and Hurtado (1996) for the Ache of Paraguay and Marlowe (2010, 137) for the Hadza of Tanzania. I am grateful to Paul Hooper for directing my attention to this information.

4. Stevenson and Wolfers (2009) present impressive cross-country and time-series questionnaire data indicating that self-reported levels of women's happiness have declined over thirty-five years relative to those of men in 125 of 147 countries examined, including the United States and the countries of the European Union. Self-reported happiness has also declined in absolute terms in the United States, though not in the European Union. For most of the paper, Stevenson and Wolfers treat these reports as equivalent to actual happiness, though at one point they acknowledge that "one might regard our rather striking observation as an opportunity to better understand the determinants of subjective well-being, and the mapping between responses to survey questions about happiness and notions of welfare" (194). This is a very much more important issue than they acknowledge. It seems likely that in the 1950s and 1960s, prior to the beginning of their US data set, there was a widespread expectation that women, especially married women, would be upbeat and positive about their condition, stifling private doubts about their own happiness in order to project a good image for the sake of their families. (Koontz [1992] is eloquent on various aspects of what she calls the "nostalgia trap" of American families.) One of the consequences of the feminist movement is that women may feel freer to express dissatisfaction; it has also

showed them that there were viable alternatives to their current lives against which those lives could be assessed. Opinions may differ about whether that was a good thing (and it's certainly conceivable that expressing dissatisfaction may increase the real levels of dissatisfaction subjects feel), but it means that we cannot interpret answers to questions about how satisfied women have been with their lives in the same way across many decades. This does not mean, of course, that the authors' conclusions are incorrect: indeed, it's possible that there is increasing female dissatisfaction fueled by the growing disparity between women's and men's levels of education. I discuss this possibility further in chapter 8.

5. Wilder, Mobasher, and Hammer 2004.

6. See, for example Tiger 1999 (*The Decline of Males*); Garcia 2008 (*The Decline of Men*); Parker 2008 (*Save the Males*); and Rosin 2010 ("The End of Men").

7. See Baumeister 2010, 17, on imprisonment and homelessness and Rosin 2010 on unemployment and educational attainment.

8. OECD 2011a.

9. See Mather and Adams 2007, especially figure 1. In addition, US Census data from 2010 corroborate the evidence from enrollment rates that increased female educational attainment, not any regression in male attainment, is the main reason for the growing gap between women's and men's attainment. See US Census Bureau 2011.

Chapter Five: Testing for Talent

1. Gowin 1915.

2. Case and Paxson 2008, 515.

3. Case and Paxson 2008, 503.

4. Case and Paxson 2008, table 4.

5. This finding is corroborated by the study reported in Abbot et al. 1998.

6. There has also been an independent finding of a correlation between height and a willingness to take risks: see Dohmen et al. 2011.

7. They do this with multiple regression analysis, which makes it possible to estimate the effect of height on earnings simultaneously with estimating the effect of talent on earnings, so that the former effect is measured taking the latter effect into account and vice versa.

8. Cinirella, Piopiunik, and Winter (2009) provide impressive evidence from Germany (where school tracking plays an important role in student educational achievement) that pupil height affects (independently of cognitive skills) the likelihood of being recommended by a primary-school teacher for enrollment in the most academic

secondary-school track. They suggest this finding is likely due to teachers' rewarding higher social skills, and they note that taller children have better social skills as early as age 3. This latter effect could be due to an innate correlation of factors leading to height and to social skills or to a feedback effect whereby adults interact more with taller children, thereby reinforcing more strongly their social skills.

9. Ogilvie 2011, 56, 60, 113, 114.

10. See Ogilvie 2003, especially introduction.

11. Greenwood et al. (2005) argue that the scale and timing of the diffusion of a large range of household labor-saving devices, from refrigerators to washing machines and vacuum cleaners, are the most plausible explanation for the increase in female labor-force participation in the US economy.

12. Goldin and Katz (2002) show that by enabling single women to delay marriage, the pill increased returns to investment in education and professional training. They suggest that it indirectly increased the attractiveness to women of delaying marriage by increasing the probability that women who married later would make better choices in the marriage market (a woman marrying later would be a relatively more attractive partner when a shift in general practices meant she was not as unusual in doing so and therefore was less likely to be stereotyped for doing so).

13. Scandinavia is an exception, because the most radical changes, such as Sweden's Marriage Law of 1915, had been in preparation before the First World War. See Therborn 2004, 74.

14. Important changes in the law included not just those relating to voting and removing formal constraints on labor market participation but also those governing the status of the spouses within a marriage, divorce, and property ownership. See Therborn 2004, especially chapter 2.

15. See Brandt 2007, 84–85.

16. Bureau of Labor Statistics 2011a.

17. Bureau of Labor Statistics 2009.

18. Bureau of Labor Statistics 2011b.

19. Bureau of Labor Statistics 2011a.

20. On board members, see Soares et al. 2010; on CEOs, see CNN Money 2011.

21. Ryan and Haslam 2005.

22. Bureau of Labor Statistics 2011b.

23. See Astur et al. 1998 on spatial navigation. Hines (2011) summarizes evidence on the influence of testosterone on a number of cognitive skills, speculating plausibly that the strong preferences of young boys for playing with mechanical toys may be due not to any particularities of shape or color of such toys but to the fact that they are designed to be pushed through space. Buss (2004, 86–87) summarizes evidence that

these differences in skill may reflect the specialization of men for hunting and women for gathering, a hypothesis that has given rise to a number of predictions about gender variations across skill types that are well supported by the evidence.

24. On verbal ability, see Hyde and Linn 1988. Hoffmann, Gneezy, and List (2011) find a gender difference in ability to solve a physical puzzle to be present in a patrilineal society but not in an adjoining matrilineal one.

25. See Fine 2011, especially chapter 3.

26. Shih, Pittinsky, and Ambady 1999. Stereotype threat has also been shown more recently to be important for responses to competition: see Günther et al. 2010.

27. Heritability refers to the degree of population variation in a trait that is explained by population genetic variation, and this is not the same thing as the total contribution of genes to the possession of a trait. For example, the number of fingers we have is almost completely determined by our genes, but its heritability is close to zero, because almost everyone has ten fingers, and those who do not have almost always lost a finger through an accident, so that the small amount of population variability is almost all due to environmental factors. Visscher et al. (2008) provide a valuable review and explanation of the notion of heritability. Estimates of the degree of heritability of IQ are sensitive to whether truly genetic contributions to IQ scores are separated from contributions from the maternal environment (even identical twins reared apart will have shared the same womb). Devlin et al. (1997) show that taking the maternal environment into account can substantially reduce estimates of the genetic heritability of IQ, from the 60–80 percent claimed by Herrnstein and Murray (1994) to somewhere between 34 percent and 48 percent by their own estimates.

28. Flynn 2007.

29. However, Blair et al. (2005) make a plausible case that it is to do with a combination of increased population access to formal schooling and (in more recent decades) to the increasing cognitive demands of mathematics education for young children, at the age when the prefrontal cortex exhibits high neural plasticity in response to experience. Ramsden et al. (2011) report neuroimaging data suggesting that changes in IQ measures in a sample of teenagers are not due to noise in the measurement of unchanging underlying skills but are correlated with modifications over time in local brain structure, specifically with gray matter in areas activated by speech (for verbal measures of IQ) and finger movements (for nonverbal measures).

30. Lynn and Irwing (2004), in a meta-analysis of fifty-seven studies using the Raven Progressive Matrices Test, find no advantage in favor of boys age 6–14 but a five-point advantage in favor of adult men, which they interpret in support of Lynn's (1999) theory that differences emerge after puberty because males mature more slowly than females. Jorm et al. (2004) found that differences on tests that favored men tended to disappear when various sociodemographic and health variables were controlled for,

but on tests that favored women they tended to be accentuated. Deary et al. (2007), in a single survey comparing pairs of opposite-sex siblings, find a very small (less than 1-point) advantage in favor of men, though they report men as having higher variance (see n. 47 below). Colom et al. (2002), on a single Spanish implementation of the Wechsler Adult Intelligence Scale, found that there were no gender differences in *g* but differences in the total IQ scores, of around 3.6 points, in favor of men, indicating that the method of aggregating component scores gives more weight to men than is warranted by calculations of *g*. Thus the fact that a test is highly *g*-loaded does not imply that differences in mean scores imply differences in the underlying *g* scores. However, Jackson and Rushton (2006), analyzing the performance of 100,000 17–18-year-olds on the Scholastic Assessment Test, find a difference in both average IQ scores and mean *g*, equivalent to 3.63 points on an IQ scale.

31. This point is also illustrated by the results reported in Geary et al. 2000, who show that a group of men performed better than a group of women on a test of arithmetical reasoning, and that individual differences in arithmetical reasoning were related to individual differences in IQ, but that there was no sex difference on the IQ test.

32. Hunt 2011, 407.

33. Lynn (1999) simply asserts that intelligence should be defined as the sum of the verbal comprehension, reasoning, and spatial abilities, which is not an argument at all. This is like claiming that a place's attractiveness to live in should be defined as the sum of its nightclubs plus the sum of its fields.

34. See Astur et al. 1998.

35. See Milner et al. 1968.

36. See Almlund et al. 2011.

37. See Miller 2009 for a general, nontechnical introduction to personality psychology, especially chapters 9 and 12–14. Schmitt et al. (2008) suggest some reasons for caution in interpreting results as stable across cultures, notably because more prosperous societies display larger differences in personality, especially between men and women.

38. Bowles and Gintis (1976) were pioneers in this area, and Bowles, Gintis, and Osborne 2001a and 2001b summarize more recent research.

39. Schmitt et al. (2008) find higher scores for women on conscientiousness in most nations in a fifty-five-nation study. However, Müller and Plug (2006) find no difference between women and men on this trait. Nyhus and Pons (2005) also find no effect of conscientiousness on earnings (and no significant difference between men and women in average conscientiousness) in a careful study based on Dutch panel data.

40. Almlund et al. 2011, figure 16, summarizes correlations with job performance, and their section 7.B discusses such findings in detail. The positive association between

emotional stability and productivity was previously found in meta-analyses by Barrick and Mount (1991) and Salgado (1997).

41. Müller and Plug 2006.

42. Almlund et al. 2011.

43. See Duckworth and Seligman 2005, for instance.

44. Müller and Plug 2006.

45. See Johnson, Carothers, and Deary 2008, 2009.

46. See Zechner et al. 2001. Miller's hypothesis is set out in Miller 2000.

47. See discussion in Hunt 2011, 382–86. Hedges and Nowell (1995, table 2) report differences in gender means and in gender variances for 37 different tests, with males having higher means in 23 of the 37 but higher variance in 35 of the 37 tests. A different selection of tests would doubtless have produced different proportions, though it seems safe to bet that the variance discrepancy would be likely to remain higher than the discrepancy in means. Deary et al. (2007) report substantially greater male than female variance in measures of *g*, sufficient to mean that roughly twice as many men and women are represented among the top 2 percent of scores.

48. See Halpern et al. 2007. Such differences are also negatively correlated across countries with other measures of gender empowerment, which supports the idea that socialization is an important influence; see Guiso et al. 2008.

49. See the various contributions to Gallagher and Kaufman 2005 for an idea of the controversy. Ellison and Swanson (2010) document a large gender gap at the highest percentiles in the United States. Although the gap does not vary much across schools, which might suggest a limited role for environmental influences, they also document that the highest-achieving girls are concentrated in a very small subset of schools. This suggests that the small degree of observed environmental variation gives little insight into the potential distribution of ability: as the authors write, "There is limited value in trying to put a lot of effort into estimating 'ability' distributions when almost all girls who would be capable of achieving extremely high scores do not do so" (29). Hyde and Mertz (2009) find that a gap does exist in the United States, though it has narrowed over time. It does not exist in some ethnic groups and in some other nations.

50. Zechner et al. 2001.

51. See Kostyniuk et al. 1996, Mayhew et al. 2003. According to a report by the Associated Press (2007), a study in 2007 by Carnegie-Mellon University for the American Automobile Association reported that there are 1.35 male deaths per 100 million miles driven, compared to 0.77 female deaths; however, I have not been able to track down the original study.

52. Gouchie and Kimura 1991; Hunt 2011, 406. See Hines 2011 for a review of the developmental effects of testosterone and resulting gender differences in behavior. However, a very recent study by Kocoska-Maras et al. (2011) finds no effect of testosterone on spatial ability in a large, double-blind, randomized study.

53. Even where physical strength matters, the discrepancies are not always what you might expect. Two of the industries in which women earn most on average, relative to men, are construction and agriculture: see Bureau of Labor Statistics 2011c.

Chapter Six: What Do Women Want?

1. An extended development of this view is set out in Browne 2002.

2. I am indebted to Bertrand 2011, which provides an excellent survey on gender differences in psychological attributes, with far more detailed discussion and references than I can include here.

3. Carey 2011.

4. These are summarized in Croson and Gneezy 2008.

5. Beyer 1990; Barber and Odean 2001; Niederle and Westerlund 2007. These gender differences are immune to the general criticisms leveled by Harris and Hahn (2011) against studies of overconfidence in general; the biases diagnosed in these studies by Harris and Hahn should not be expected to operate differently for women and men.

6. This was first shown by Bateman (1948); see also Trivers 1972, especially 37–39.

7. For example, Sukumar and Gadgil (1988) show that there are differences in risk-taking behavior between male and female Asian elephants, with males tending to take greater risks.

8. Bertrand and Mullainathan (2010) provide strong evidence that CEOs are rewarded as much for the results of luck as for anything that might represent the influence of their own efforts on firm performance, and that this tendency is stronger in firms that, by other criteria, appear to be less well governed (less in the interests of shareholders, that is). They interpret this finding, plausibly, as reflecting the fact that CEOs try to set their own pay and have as much interest in rewarding themselves for the results of luck as for the results of their efforts. This leaves a number of questions unanswered, notably why even poorly governed firms do not resist payment for luck more strongly than payment for effort; they clearly must offer some resistance, as CEOs cannot choose to pay themselves any amount they like.

9. Scotchmer (2008) develops a model of promotion in hierarchies that formalizes this insight.

10. Andreoni and Vesterlund 2001. Engel (2011) finds in a meta-analysis of dictator games that women systematically behave more altruistically than men. However, this is far from being an agreed conclusion. Baumeister (2010, especially chapter 5) argues that men and women tend to express altruism in different settings, with women expressing it preferentially in small groups and within intimate relationships, and men expressing it in larger groups and more public settings.

11. See Chaudhuri 2011. Pinker (2011, 684–89) speculates that overall levels of violence in society might be influenced by the extent to which women hold positions of power.

12. Niederle and Westerlund 2007. Interestingly, even the women in the highest quartile of performance were less likely to choose the tournament than the men in the lowest quartile.

13. Kuhn and Villeval 2011.

14. Gneezy, Niederle, and Rustichini 2003; Niederle and Westerlund 2008. Booth and Nolen (2009) find that girls from single-sex schools compete as enthusiastically as boys, suggesting social learning and not genetics as the reason behind observed gender differences in competitive behavior. These studies are limited to school-age children, and the findings may not generalize to other age groups.

15. Kuhn and Villeval 2011.

16. See Gupta, Poulsen, and Villeval 2005 for an example of a study where it does not make a difference.

17. Gneezy and Rustichini (2004) found that girls competed more strongly against boys than against other girls in a test of running.

18. See Dreber, Von Essen, and Ranehill 2011 for a study on Swedish children that fails to find the differences previously found for Israeli children by Gneezy and Rustichini 2004. Cárdenas et al. (2011) find no gender differences in competitiveness in Colombia, whereas in Sweden the results are mixed depending on the task studied.

19. Günther et al. 2010.

20. See Chen, Katusczak, and Ozdenoren 2009; Apicella et al. 2008; Dreber et al. 2009; Dreber and Hoffman 2010. However Apicella et al. (2011) find no hormonal correlates of choosing whether to compete in solving mazes.

21. Manning and Swaffield 2008; Manning and Saidi 2010. Booth 2009 gives an excellent overview of the issues.

22. Babcock and Laschever 2003, 7.

23. Bowles, Babcock, and McGinn 2005; Small et al. 2007.

24. Bowles, Babcock, and Lai 2007.

25. This is a point made long ago about discrimination by Gary Becker (1971).

26. See Goldacre 2009, chapter 4.

27. This point is made very effectively by Dougherty (2005).

28. Bertrand, Goldin, and Katz 2010.

29. This doesn't prove that there's no difference in talent: there might be certain characteristics of women who choose to become mothers that make them less likely to be successful some years later. But any such difference would have had to be invisible to employers before the women's decision to take career breaks, so in the absence of more concrete evidence, the career breaks themselves look like a more plausible explanation for the divergence in salaries.

30. Bertrand, Goldin, and Katz 2010, 240.
31. Yoshino 2006, ix.

Chapter Seven: Coalitions of the Willing

1. de Waal 1989, 48.
2. de Waal 1989, 51.
3. Goodall 1986; Nishida 1996.
4. de Waal 1989, 122.
5. Low 2000, 181.
6. On primates, see Henazi and Barrett 1999; Silk, Alberts, and Altmann 2004; Pandit and van Schaik 2003; van Schaik, Pandit, and Vogel 2004, 2005. On other species, see Low 2000, 181–82.
7. Granovetter 1973.
8. It's quite possible that common membership in a group may by itself determine willingness to treat one another favorably. If men treat other men more favorably than they treat women just because they are men, this would be an example of classic discrimination, which may indeed be part of the reason why women appear disadvantaged in the labor market. However, such discrimination is illegal in many countries (except when recruiting for a few occupations, such as military ones). I am interested here in a more indirect and subtle process whereby more favorable treatment is determined by membership of a coalition in which the members have invested some effort in building bilateral ties.
9. Granovetter 1973.
10. Moore 1990.
11. A study by Chow and Ng (2007) finds that among coworkers, women are less likely than men to socialize with those whom they subsequently need to approach for instrumental favors, though the sample was small and far from random (seventy-two executive-education participants, of whom two-thirds were men). A study by Forret and Dougherty (2004), reporting a tendency for women's networking behavior to be less effective at promoting their professional interests than that of men, indicates a small negative correlation between female gender and three main measures of networking activity, namely "maintaining external contacts," "socializing," and "engaging in professional activities," though it does not indicate the significance of this correlation. Campbell (1988) reports that in a sample of employed people changing jobs, women had contacts in a smaller range of occupations than did men.
12. Ibarra 1997.
13. Bu and Roy 2005.
14. Burt 1998.

15. Green and Singleton 2009; Igarashi, Takai, and Yoshida 2005; Lemish and Cohen 2005.

16. Two studies that have done so are Smoreda and Licoppe, 2000 and Wajman, Bittman, and Brown 2009.

17. Friebel and Seabright 2011.

18. Tannen 1990 and 1994 are studies of gender differences in conversational strategies (the latter focusing on the workplace), though based on examples rather than statistical evidence.

19. Forret and Dougherty (2004) find that self-reports of networking activities along several dimensions are sometimes insignificantly related with measures of professional success for women even when they are positively related for men, though this finding is not consistent across different measures. Aguilera (2008) reports that women who find employment through personal contacts do significantly better than those who find employment through ordinary labor market procedures, while this effect is not present for men; however, this result is compatible with the possibility that men's employment is already more remunerative because of male networking, whether or not the job is found through personal contacts.

20. We also have information on educational links and on connections via membership of not-for-profit associations.

21. Lalanne and Seabright 2011.

22. The basic difficulty is that we cannot measure talent directly, so we need to find a proxy for talent. If we use as explanatory variables in our regression analyses some other variable that is imperfectly correlated with talent, it will not take out of our measured effect enough of the hidden effect of talent, so our estimates will still be biased in the direction of finding more of an effect than really exists. Alternatively, if we use past values of salaries as proxies for talent, this has the opposite bias: namely that we risk overlooking an effect where one really exists, because any effect of networks on past salaries will be excluded from our analysis by definition. We use two variants of this second approach to tackle the problem, and fortunately both methods give comparable results. One is to use salaries several years in the past as control variables so as to allow several years for the effects of networks on salaries to manifest themselves. This is particularly important because the contacts people make in one year may not benefit their careers until several years later. The other variant is to look just at what happens when people's networks change over time: do increases in a given person's network lead to increases in their salaries? For further details, see Lalanne and Seabright 2011.

23. This part of the analysis is based on joint work with Nicoletta Berardi of the Banque de France, reported in Berardi and Seabright 2011.

24. Other researchers have found that gender differences in deferred remuneration are even greater than those in salaries: see Albanesi and Olivetti 2008; Kulich et al. 2009; Bebchuk and Fried 2004; and Geletkanycz, Boyd, and Finkelstein 2001.

25. This is not a new concern: Daily, Certo, and Dalton (1999) made much the same point when reviewing evidence of female boardroom representation in the 1990s. A recent report in the *Observer* (Bawden 2011) suggested that firms in the United Kingdom were focusing only on nonexecutive and not on executive board appointments, in response to the Davies report that recommended a voluntary target of 25 percent female boardroom representation. A related concern is expressed by Gregoric et al. (2010), who present evidence suggesting that boards with higher representation of women tend to be less diverse in other dimensions.

26. See Hunt 2010, 97.

27. There is indeed a literature discussing which of these motivations is the more important, with Geletkanycz, Boyd, and Finkelstein (2001) favoring (a particular version of) the prudence theory and Bebchuk and Fried (2004) favoring a version of the preference theory.

28. Low 2000, chapter 10.

29. Scott (1983, 319) reports that "respondents reporting to someone of the same gender had significantly higher trust in their superior than did men or women reporting to a superior of the opposite sex." Unfortunately, this finding does not show that their trust was warranted. And it is directly contradicted by a study by Jeanquart-Barone (1993), which found that "the highest level of trust was found between female subordinates reporting to male supervisors" (1).

30. Bonein and Serra (2009) find evidence in favor of greater trust between players of the same gender, while Sutter et al. (2009) find evidence against it.

Chapter Eight: The Scarcity of Charm

1. Giridharadas 2007.

2. Giraudoux 1997, 32.

3. Mayhew 1861, 1: 311–12.

4. Nasscom 2009.

5. On Africa, see GSMA/A. T. Kearney 2011a, which reports 620 million mobile subscriptions in the top twenty-five African countries (p. 4); on India and China, see GSMA/A. T. Kearney 2011b, which reports 752 million subscriptions in India and 842 million in China (p. 1).

6. GSMA/A. T. Kearney, 2011b, 4.

7. International Telecommunication Union, 2011a for Internet subscriptions, 2011b for Internet users.

8. There have been more than 52 million visits logged for Khan Academy videos, though how many visitors have fully used the courses is a matter of conjecture. See www.khanacademy.org.

9. This is a potentially vast subject that in principle would require another book. Interesting theoretical papers from an economic point of view include Falkinger 2007 and 2008, and Anderson and De Palma 2009. Klingberg 2009 gives a very accessible account of the neuroscientific constraints on human attention and their consequences for our everyday lives.

10. Bureau of Labor Statistics 2010.

11. The data on occupational earnings discussed in the next few paragraphs are all available at Bureau of Labor Statistics, 2011d; I have drawn on the series "May 2009 National Occupational Employment and Wage Estimates" and "2000 National Occupational Employment and Wage Estimates."

12. More precisely, an ordinary least-square regression analysis (over the 686 occupations for which there are data for the two years) of the ratio of the earnings spread in 2009 to the earnings spread in 2000, with the percentage change in annual median earnings and the change in total employment as regressors, yields coefficients of 8.9 percent on the first and 2.4 percent on the second; each coefficient is statistically significant at a little under 2 percent.

13. Farrell 1993.

14. OECD 2011b.

15. See US Census Bureau 2011.

16. For marriage rates among whites and African Americans, see "Down or Out" 2011. For differences in educational attainment, see US Census Bureau 2011. Among the population aged 30–34, 27.4 percent of black women have at least a bachelor's degree, while only 16.9 percent of black men do. Banks (2011) argues that this disparity is creating a real crisis in relationships for both black men and black women.

17. Stevenson and Wolfers 2009.

18. The authors also report strong positive trends for the reported happiness of African American women. There are in any case some difficulties in interpreting trends in reported happiness over time. See the introduction to part two, n. 4, for a discussion of the issues involved.

Chapter Nine: The Tender War

1. The ape anthropologists would of course acknowledge that some elements of their story had been anticipated by the work of human anthropologists before them. Hrdy 2009 and Kaplan et al. 2009 summarize the literature on the first of these points. Hrdy's book begins with an intriguing thought experiment imagining what would happen if her fellow passengers on a plane journey were other species of ape. "What if I were traveling with a planeload of chimpanzees? Any one of us would be lucky to dis-

embark with all ten fingers and toes still attached, with the baby still breathing and unmaimed. Bloody earlobes and other appendages would litter the aisles" (3). Bowles and Gintis (2011) discuss how the evolution of prosocial behavior for elaborate cooperative purposes has driven the development of most of the other characteristics that distinguish humans from the other great apes. Burkart, Hrdy, and van Schaik (2009) propose a theory of the evolution of the motivational components of the distinctively human psychological complex based on cooperative breeding; chimpanzees, they argue, have many of the purely cognitive components already in place.

2. The literature on this topic is surveyed in Seabright 2010, especially chapters 3–5. Pinker 2011 has a comprehensive overview of the literature on the downward trend of violence in recorded history.

3. This point is made most forcefully in Boehm 1999.

4. See, for instance, de Waal 2001, especially chapter 2.

5. This is a central theme, perhaps the principal theme, of Foucault 1990.

6. The public interest in the affair allegedly centered on whether Clinton had committed perjury in his statements about Lewinsky in relation to the lawsuit brought against him by Paula Jones for sexual harassment. But it would be hard for an objective observer to deny that the press and public fascination with presidential adultery became overwhelming. Otherwise it is hard to see why the allegations of a consensual relationship with Lewinsky should have been considered relevant to allegations of harassment of Jones or why Clinton should have felt such a strong incentive to lie under oath about the former.

7. Educated and reflective people are by no means immune from such reactions. For example, in a discussion of sexual morality, the philosopher Elizabeth Anscombe wrote of "the rewardless trouble of the spirit associated with the sort of sexual activity which from its type is guaranteed sterile: the solitary or again the homosexual sort" (1976). This led her commentators in the same volume, Michael Tanner and Bernard Williams (1976), to inquire, with reference to homosexual acts, "How does Professor Anscombe claim to know?"

8. See, for example, Marlowe 2010, 175.

9. Several chapters in Muller and Wrangham 2009 discuss sexual coercion in humans and the insights that emerge from comparing it with coercion in nonhuman primates; the final chapter, in particular, draws general lessons from the many case studies in the volume.

10. See Muller and Wrangham 2009, chapter 18.

11. This has been an important issue in the development of hostility between radical Islam and modern industrialized societies. Sayyid Qutb, the Egyptian founder of the Muslim Brotherhood and an important influence on Osama bin Laden and the 9/11 hijackers, was significantly influenced by a visit to the United States between 1948 and

1950. In his best-selling work *Milestones*, published two years before his execution by Gamal Abdel Nasser's regime, Qutb railed against what he called "this behavior, like animals, which you call 'free mixing of the sexes.' This vulgarity which you call emancipation of women" (Qutb 1964, 139). He also described churches as "sexual playgrounds" (cited in Irwin 2001) and wrote that "the American girl is well acquainted with her body's seductive capacity. She knows seductiveness lies in the round breasts, the full buttocks, and in the shapely thighs, sleek legs and she shows all this and does not hide it" (cited in von Drehle 2006).

12. Fischer 1994 is a detailed social history of the telephone in America up to and including the 1930s.

13. Besides the poster reproduced in the figure, others can be viewed at the *Sodahead* blog, http://sodahead.com/living/when-smoking-was-good-for-your-health-socially-acceptable/blog-63461, accessed June 12, 2011.

14. Stephanie Coontz's book *Marriage: A History* (2005), which documents this development in great detail, has the subtitle *From Obedience to Intimacy, or How Love Conquered Marriage.*

15. There is no doubt that the physical manifestations of erotic passion can be understood through the adaptive logic of natural selection (see Fisher 2004 for a detailed account and Young 2009 for a summary of some of the scientific contributions to this literature; Frazzetto 2010 emphasizes that even online dating, with its entirely artificial setting, can be usefully illuminated using the same explanatory tools). But this does not mean that a romantic relationship can be best understood by analogy with an organ of the body.

16. DeSteno, Vadesolo, and Bartlett (2006), for instance, report experimental evidence suggesting that jealousy reactions are prompted by threats to self-esteem, which implies (though it does not prove) that contextual variation in the extent of threats to self-esteem is likely to affect the extent to which jealousy is felt in otherwise similar relationship dynamics. Hupka and Ryan (1990) report that manifestations of aggressive male responses in situations eliciting jealousy are substantially variable across a sample of ninety-two preindustrial societies in response to the presence or absence of a number of stable cultural norms, but that female responses do not vary significantly with such norms.

17. Wasson 2010, 127–29. Rather less subtly, an advertisement for a fashion brand in the free Parisian commuter magazine *20 Minutes* had the slogan "Elle change de look comme elle change de boyfriend" (March 2, 2011).

18. Though space precludes discussion of this point in detail here, the same applies to relationships that differ in other ways from the model. It is common, for instance, for critics to decry commercial sex as sordid and tawdry (see Jeffreys 1997, for instance)

and to deny that it represents a real choice for the seller. What matters is not how it compares with an ideal consensual encounter based on mutual desire but rather how it compares with the actual alternatives available to the parties (though child prostitution raises issues that go far beyond those relevant to commercial sex between adults). Unfortunately here again, legal or social constraints on consensual encounters make it more difficult to take action against coerced encounters: a study of street-level prostitution in Chicago in 2007 showed that a prostitute was more likely to have given free sexual services to a police officer than to have been arrested by one (Levitt and Venkatesh 2007). It is also likely that criminalizing commercial encounters between consenting adults discourages cooperation with the police by some of the most likely witnesses of sexual crimes against nonconsenting adults, or against children.

Many ethnographies have emphasized the variety of motivations and circumstances surrounding prostitution, both on the side of clients and on the side of the prostitutes themselves. See Clouet 2008 for a recent example; Meston and Buss 2009, chapter 8, for evidence based on an online survey; and Zelizer 2006 for many examples and a comprehensive discussion. Ringdal 1997 is a detailed historical account of the way the institutions of prostitution have varied across many different societies. Edlund and Korn (2002) have shown that many aspects of commercial sex are consistent with the idea that prostitutes undertake it as an economic choice: it is not one that many readers would envy, perhaps, but the alternatives available to many prostitutes are not enviable either.

Edlund and Korn emphasize, though, that the short-term attractions of female prostitution typically come with a long-term cost, namely the destruction of the marriage prospects of the women concerned. This cost is evidently much higher for street prostitutes than for others: indeed, many of the part-time student prostitutes described in Clouet 2008 are unlikely ever to reveal their former occupation to a future husband, who is unlikely ever to find out unless they are careless. And it hardly needs repeating that in cultures all over the world, courtship may involve substantial in-kind investments by one party (usually though not always male) as inducements for the sexual favors of the other, even when no money changes hands, so that the boundaries between commercial and noncommercial sex can be very blurred. This does not mean, as Zelizer (2006) emphasizes, that courtship is "nothing but" commercial sex: on the contrary, each type of exchange comes freighted with complex attitudes, norms, and expectations that distinguish it from others. But nor are these wholly distinct realms with nothing in common: they have a family resemblance, one might say. What distinguishes commercial sex may be the short-term nature and explicit character of the transaction rather than the fact of the transaction per se. In a different market in which unease has often been expressed for similar reasons about commercial transactions, namely the market for human gametes, a study by Almeling (2011), based on detailed interviews, emphasizes that differences in norms and conventions mean that "sperm donation is considered a job and egg donation a gift" in spite of the fact that both are remunerated in the United States (168).

19. There is evidence that job satisfaction is more highly correlated with reported life satisfaction for men than for women (Della Giusta, Jewell, and Kambhampati 2011), but evidently this finding does not tell us to what extent the correlation arises because of inheritance rather than social reinforcement. It's possible that there are genetic foundations for a degree of variability in the single-mindedness with which individuals focus on their jobs to the exclusion of other aspects of their lives, consistent, for instance, with the observation that autism is more common among men than among women (see Baron-Cohen 2003). But it is also my (unscientific) impression that very single-minded men are much more likely to be profiled admiringly in the press than are equivalently single-minded women. The most admiring press profiles are reserved for women who manage to "have it all," whereas profiles of those who pursue single goals (at least professional goals) with great concentration often carry a hint of monstrosity, with the partial exception of sportswomen whose careers finish early enough for them to switch their focus to other goals, including a family or a second career. If my impression is accurate (and I would be interested to see a careful study that tested it), it would imply that there is an important contribution of social conditioning to the tendency of men to pursue less diverse goals.

20. Apter 1995.

21. Fernández and Cheng Wong (2011) report a large difference in education and labor-force participation between women born in 1935 and those born twenty years later.

22. US Census Bureau 2011. Women in their late forties and early fifties are slightly more likely to be educated to bachelor's degree level and above than men in the same cohort. In Lalanne and Seabright 2011, we report that women in our sample are slightly more educated than men. However, the composition of degrees may also matter for remuneration, and men are somewhat more likely to have degrees in finance and in engineering.

23. Systematic evidence on how this has worked in the few countries that have tried it is scarce. See Bennhold 2010. Existing parental-leave provisions are highly asymmetric between parents: for regularly updated international comparisons, see International Labor Organization 2011.

24. See Cools, Fiva, and Kirkebøen (2011), who study the effect of a reform in Norway in 1993 that substantially increased the incentives for fathers to take paternity leave. They find an essentially negligible impact of the provision on a range of variables, including average school performance, family fertility, and divorce rates (though these

findings may reflect the rather minor nature of the reform). However, children's school performance became more responsive than before the reform to the education levels of the father, suggesting that fathers are becoming more involved in care of children. They find that maternal labor supply is complementary to paternal labor supply, so in families where fathers take time off work to care for children, mothers are more rather than less likely to reduce working hours and earnings.

REFERENCES

Abbott, Robert D., Lon R. White, G. Webster Ross, Helen Petrovitch, Kamal H. Masaki, David A. Snowdon, and J. David Curb. 1998. "Height as a Marker of Childhood Development and Late Life Cognitive Function: The Honolulu-Asia Ageing Study." *Pediatrics* 102: 602–9. doi:10.1542/peds.102.3.602.

Acton, William. [1857] 2009. *The Functions and Disorders of the Reproductive Organs in Childhood, Youth, Adult Age and Advanced Life.* Facsimile ed. Charleston, SC: BiblioLife.

Adovasio, J. M., Olga Soffer, and Jake Page. 2007. *The Invisible Sex: Uncovering the True Roles of Women in Prehistory.* New York: HarperCollins.

Aguilera, Michael Bernabé. 2008. "Personal Networks and the Incomes of Men and Women in the United States: Do Personal Networks Provide Higher Returns for Men or Women?" *Research in Social Stratification and Mobility* 26(3): 221–33.

Albanesi, Stefania, and Claudia Olivetti. 2008. "Gender and Dynamic Agency: Theory and Evidence on the Compensation of top executives." Unpublished manuscript.

Almeling, Rene. 2011. *Sex Cells: The Medical Market for Eggs and Sperm.* Berkeley: University of California Press.

Almlund, Mathilde, Angela Lee Duckworth, James Heckman, and Tim Kautz. 2011. *Personality Psychology and Economics.* Discussion Paper 5500, Institute for the Study of Labor, Bonn.

Alvarez, Helen. 2004. "Residence Groups among Hunter-Gatherers: A View of the Claims and Evidence for Patrilocal Bands." In *Kinship and Behavior in Primates*, ed. Bernard Chapais and Carol M. Berman. New York: Oxford University Press.

Anderson, Simon, and André de Palma. 2009. "Information Congestion." *Rand Journal of Economics* 40: 688–709.

Andreoni, James, and Lise Vesterlund. 2001. "Which Is the Fair Sex? Gender Differences in Altruism." *Quarterly Journal of Economics* 116: 293–312.

Anscombe, E. L. 1976. "Contraception and Chastity." In *Ethics and Population*, ed. M. D. Bayles. Cambridge, MA: Schenkman.

Apicella, C. L., A. Dreber, B. C. Campbell, P. B. Gray, M. Hoffman, and Anthony C. Little. 2008. "Testosterone and Financial Risk Preferences." *Evolution and Human Behavior* 29(6): 384–90.

Apicella, Coren L., Anna Dreber, Peter B. Gray, Moshe Hoffman, Anthony C. Little, and Benjamin C. Campbell. 2011. "Androgens and Competitiveness in Men." *Journal of Neuroscience, Psychology and Economics* 4(1): 54–62.

Apter, Terri. 1995. *Working Women Don't Have Wives*. London: Palgrave Macmillan.

Ariely, Dan. 2008. *Predictably Irrational*. New York: HarperCollins.

Arnqvist, Göran, and Locke Rowe. 2005. *Sexual Conflict*. Princeton, NJ: Princeton University Press.

Associated Press. 2007. "Bad Female Drivers? It's Just a Myth, New Analysis Finds." *St. Petersburg Times*. January 19. www.sptimes.com/2007/01/19/Worldandnation/Bad_female_drivers_It.shtml.

Astur, Robert S., Maria L Ortiz, and Robert J. Sutherland. 1998. "A Characterization of Performance by Men and Women in a Virtual Morris Water Task: A Large and Reliable Sex Difference." *Behavioural Brain Research* 93: 185–90.

Aureli, Filippo, and Frans B. M. de Waal, eds. 2000. *Natural Conflict Resolution*. Berkeley: University of California Press.

Babcock, Linda, and Sara Laschever. 2003. *Women Don't Ask: Negotiation and the Gender Divide*. Princeton, NJ: Princeton University Press.

Banks, Ralph Richard. 2011. *Is Marriage for White People? How the African American Marriage Decline Affects Everyone*. New York: Dutton Adult.

Barash, David P., and Judith Eve Lipton. 2001. *The Myth of Monogamy: Fidelity and Infidelity in Animals and People*. New York: Henry Holt.

Barber, Brad M., and Terrance Odean. 2001. "Boys Will Be Boys: Gender, Overconfidence, and Common Stock Investment." *Quarterly Journal of Economics* 116: 261–92.

Barber, Elizabeth Wayland. 1994. *Women's Work: The First 20,000 Years*. New York: W. W. Norton.

Baron-Cohen, Simon. 2003. *The Essential Difference: The Truth about the Male and Female Brain*. New York: Basic Books.

Barrick, M. R., and M. K. Mount, 1991. "The Big Five Personality Dimensions and Job Performance: A Meta-analysis." *Personnel Psychology* 44: 1–26.

Bateman, A. J. 1948. "Intrasexual Selection in Drosophila." *Heredity* 2: 349–68.

Baumeister, Roy F. 2010. *Is There Anything Good about Men? How Cultures Flourish by Exploiting Men*. Oxford: Oxford University Press.

Baumeister, Roy F., and Kathleen D. Vohs. 2004. "Sexual Economics: Sex as Female Resource for Social Exchange in Heterosexual Interactions." *Journal of Personality and Social Psychology* 8: 339–63.

Bawden, Tom. 2011. "Surge in Appointments of Female Board Members Shows Companies Heeding Compulsory Quota Threat." *Guardian*. August 21. http://guardian.co.uk/business/2011/aug/21/fears-quotas-more-women-boardroom.

Bebchuk, Lucian, and Jesse Fried. 2004. *Pay without Performance: The Unfulfilled Promise of Executive Compensation*. Cambridge, MA: Harvard University Press.

Becker, Gary. 1971. *The Economics of Discrimination*. 2nd ed. Chicago: University of Chicago Press.

Bénabou, Roland, and Jean Tirole. 2006. "Incentives and Pro-social Behavior." *American Economic Review* 96: 1652–78.

Bennhold, Katrin. 2010. "In Sweden, Men Can Have it All." *New York Times*. June 9. http://nytimes.com/2010/06/10/world/europe/10iht-sweden.html.

Berardi, Nicoletta, and Paul Seabright. 2011. *Network and Career Coevolution*. Discussion Paper 8632, Centre for Economic Policy Research, London.

Berggren, Niclas, Henrik Jordahl, and Panu Poutvaara. 2006. "The Looks of a Winner: Beauty and Electoral Success." *Journal of Public Economics* 94: 8–15.

Berman, Louis A. 2003. *The Puzzle: Exploring the Evolutionary Puzzle of Male Homosexuality*. Wilmette, IL: Godot.

Bertrand, Marianne. 2011. "New Perspectives on Gender." In *Handbook of Labor Economics*, vol. 4b, ed. Orley Ashenfelter and David Card. Amsterdam: North-Holland.

Bertrand, Marianne, and Sendhil Mullainathan. 2001. "Are CEOs Rewarded for Luck? The Ones without Principals Are." *Quarterly Journal of Economics* 118: 901–32.

Bertrand, Marianne, Claudia Goldin, and Lawrence F. Katz. 2010. "Dynamics of the Gender Gap for Young Professionals in the Financial and Corporate Sectors." *American Economic Journal: Applied Economics* 2: 228–55. www.aeaweb.org/articles.php?doi:10.1257/app.2.3.228.

Beyer, Sylvia. 1990. "Gender Differences in the Accuracy of Self-Evaluations of Performance." *Journal of Personality and Social Psychology* 59: 960–70.

Bilimoria, D., and S. K. Piderit. 1994. "Board Committee Membership: Effects of Sex-Based Bias." *Academy of Management Journal* 37(6): 1453–77.

Birkhead, Tim. 2000. *Promiscuity*. Cambridge, MA: Harvard University Press.

Blair, Clancy, David Gamson, Steven Thorne, and David Baker. 2005. "Rising Mean IQ: Cognitive Demand of Mathematics Education for Young Children, Population Exposure to Formal Schooling, and the Neurobiology of the Prefrontal Cortex." *Intelligence* 33: 93–106.

Boehm, Christopher. 1999. *Hierarchy in the Forest: The Evolution of Egalitarian Behavior*. Cambridge, MA: Harvard University Press.

Bond, J. R., and W. E. Vinacke. 1961. "Coalitions in Mixed-Sex Triads." *Sociometry* 24: 61–75.

Bonein, Aurélie, and Daniel Serra. 2009. "Gender Pairing Bias in Trustworthiness." *Journal of Socio-economics* 38: 779–89.

Booth, Alison. 2009. *Gender and Competition.* Discussion Paper 4300, Institute for the Study of Labor, Bonn.

Booth, Alison, and Patrick Nolen. 2009. "Choosing to Compete: How Different Are Girls and Boys?" Discussion Paper 7214, Centre for Economic Policy Research, London.

Borgerhoff Mulder, Monique. 1992. "Reproductive Decisions." In *Evolutionary Ecology and Human Behaviour,* ed. Alden Smith and B. Winterhalder. New York: Aldine de Gruyter.

Bowlby, John. 1988. *A Secure Base: Clinical Applications of Attachment Theory.* London: Routledge.

Bowles, Hannah R., Linda Babcock, and Lei Lai. 2007. "Social Incentives for Sex Differences in the Propensity to Initiate Negotiation: Sometimes It Does Hurt to Ask." *Organizational Behavior and Human Decision Processes* 103: 84–103.

Bowles, Hannah R., Linda Babcock, and Kathleen L. McGinn. 2005. "Constraints and Triggers: Situational Mechanics of Gender in Negotiation." *Journal of Personality and Social Psychology* 89: 951–65.

Bowles, Samuel. 2009. "Did Warfare among Ancestral Hunter-Gatherer Groups Affect the Evolution of Human Social Behaviors?" *Science* 324: 1293–98.

Bowles, Samuel, and Jung-Kyoo Choi. 2007. "The Coevolution of Parochial Altruism and War." *Science* 318: 636–40.

Bowles, Samuel, and Herbert Gintis. 1976. *Schooling in Capitalist America: Educational Reform and the Contradictions of Economic Life.* New York: Basic Books.

———. 2011. *A Cooperative Species: Human Reciprocity and Its Evolution.* Princeton, NJ: Princeton University Press.

Bowles, Samuel, Herbert Gintis, and Melissa Osborne. 2001a. "The Determinants of Earnings: A Behavioral Approach." *Journal of Economic Literature* 39: 1137–76.

———. 2001b. "Incentive-Enhancing Preferences: Personality, Behavior, and Earnings." *American Economic Review* 91: 155–58.

Boyd, Brian. 2012. *Why Lyrics Last: Evolution, Cognition and Shakespeare's Sonnets.* Cambridge, MA: Harvard University Press.

Brandt, Allan M. 2007. *The Cigarette Century.* New York: Basic Books.

Brown, W. M., L. Cronk, K. Grochow, A. Jacobson, K. Liu, Z. Popovic, and R. Trivers. 2005. "Dance Reveals Symmetry Especially in Young Men." *Nature* 438(22): 1148–50.

Browne, Kingsley R. 2002. *Biology at Work: Rethinking Sexual Equality.* Brunswick, NJ: Rutgers University Press.

Bu, Nailin, and Jean-Paul Roy. 2005. "Career Success Networks in China: Sex Differences in Network Composition and Social Exchange Practices." *Asia Pacific Journal of Management* 22(4): 381.

Bureau of Labor Statistics. 2009. *Labor Force Participation of Women and Mothers, 2008.* www.bls.gov/opub/ted/2009/ted_20091009_data.htm#a.

———. 2010. *Occupations with Similar Medians, but Differing Wage Variation, 2008.* www.bls.gov/oes/wage_discussions_table1.htm.

———. 2011a. *Women at Work.* www.bls.gov/spotlight/2011/women/.

———. 2011b. *Employed Persons by Detailed Occupation, Sex, Race, and Hispanic or Latino Ethnicity.* www.bls.gov/cps/cpsaat11.pdf.

———. 2011c. *Women's Employment and Earnings by Industry, 2009.* www.bls.gov/opub/ted/2011/ted_20110216_data.htm.

———. 2011d. *Occupational Employment Statistics.* www.bls.gov/oes/oes_arch.htm.

Burkart, J. M., S. B. Hrdy, and C. M. van Schaik. 2009. "Cooperative Breeding and Human Cognitive Evolution." *Evolutionary Anthropology* 18: 175–86.

Burt, Ronald S. 1998. "The Gender of Social Capital." *Rationality and Society* 10: 5–46.

Buss, David M. 2003. *The Evolution of Desire: Strategies of Human Mating.* New York: Basic Books.

———. 2004. *Evolutionary Psychology: The New Science of the Mind.* Boston: Pearson Education.

Caldwell, H. K., and W. S. Young. 2006. "Oxytocin and Vasopressin: Genetics and Behavioral Implications." In *Handbook of Neurochemistry and Molecular Neurobiology: Neuroactive Proteins and Peptides.* 3rd ed., ed. A. Lajtha and R. Lim, 573–607. Berlin: Springer.

Campbell, Bernard G., ed. 1972. *Sexual Selection and the Descent of Man.* Chicago: Aldine.

Campbell, Karen E. 1988. "Gender Differences in Job-Related Networks." *Work and Occupations* 15: 179–200.

Cárdenas, Juan-Camilo, Anna Dreber, Emma von Essen, and Eva Ranehill. 2011. "Gender Differences in Competitiveness and Risk Taking: Comparing Children in Colombia and Sweden." *Journal of Economic Behavior and Organization*, doi:10.1016/j.jebo.2011.06.008.

Carey, Benedict. 2011. "Need Therapy? A Good Man Is Hard to Find." *New York Times.* May 21.

Carroll, Sean. 2010. "The Making of the Fittest: The DNA Record of Evolution." In *Darwin*, ed. William Brown and Andrew C. Fabian. Cambridge: Cambridge University Press.

Case, Anne, and Christina Paxson. 2008. "Stature and Status: Height, Ability and Labor Market Outcomes." *Journal of Political Economy* 116: 499–532.

Centorrino, Samuele, Elodie Djemai, Astrid Hopfensitz, Manfred Milinski, and Paul Seabright. 2011. *Smiling Is a Costly Signal of Cooperation Opportunities: Experi-*

mental Evidence from a Trust Game. Discussion Paper 8374, Centre for Economic Policy Research, London.

Chapais, Bernard, and Carol M. Berman, eds. 2004. *Kinship and Behavior in Primates*. New York: Oxford University Press.

Chaudhuri, Ananish. 2011. "Gender and Corruption: A Survey of the Evidence." Unpublished manuscript.

Chen, Y., P. Katusczak, and E. Ozdenoren. 2009. "Why Can't a Woman Bid More Like a Man?" Working Paper 275, Center for Economic Research and Graduate Education, Economics Institute, Prague.

Chow, I. H., and I. Ng. 2007. "Does the Gender of the Manager Affect Who He/She Networks With?" *Journal of Applied Business Research* 23: 49–60.

Clouet, Eva. 2008. *La prostitution étudiante*. Paris: Max Milo.

CNN Money. 2011. "Fortune 500: Women CEOs." http://money.cnn.com/magazines/fortune/fortune500/2011/womenceos/.

Cochrane, Gregory, and Henry Harpending. 2009. *The 10,000-Year Explosion: How Civilization Accelerated Human Evolution*. New York: Basic Books.

Colom, Roberto, Luis Garcia, Manuel Juan-Espinosa, and Francisco Abad. 2002. "Null Sex Differences in General Intelligence: Evidence from the WAIS-III." *Spanish Journal of Psychology* 5: 29–35.

Cools, Sara, Jon H. Fiva, and Lars Johannessen Kirkebøen. 2011. "Causal Effects of Paternity Leave on Children and Parents." Discussion Paper 657, Research Department, Statistics Norway, Oslo.

Coontz, Stephanie. 1992. *The Way We Never Were: American Families and the Nostalgia Trap*. New York: Basic Books.

Cooper, Michael. 2008. "No. 1 Faux Pas in Washington? Candor, Perhaps." *New York Times*. June 25.

Copeland, S. R., M. Sponheimer, D. J. de Ruiter, J. A. Lee-Thorp, D. Codron, P. J. le Roux, V. Grimes, and M. P. Richards. 2011. "Strontium Isotope Evidence for Landscape Use by Early Hominins." *Nature* 474: 76–78.

Cornwallis, Charlie K., Stuart A. West, Katie E. Dais, and Ashleigh S. Griffin. 2010. "Promiscuity and the Evolutionary Transition to Complex Societies." *Nature* 466: 969–72. doi:10.1038/nature09335.

Coyne, Jerry. 2009. *Why Evolution Is True*. New York: Penguin.

Crawley, Ernest. 1929. *Studies of Savages and Sex*. Edited by Theodore Besterman. London: Methuen.

Cronk, L., N. Chagnon, and W. Irons, eds. 2000. *Adaptation and Human Behaviour: An Anthropological Perspective*. New York: Aldine de Gruyter.

Croson, R., and U. Gneezy. 2009. "Gender Differences in Preferences." *Journal of Economic Literature* 47(2): 1–27.

Crowell, Judith, and Everett Waters. 2005. "Attachment Representations, Secure-Base Behavior and the Evolution of Adult Relationships." In *Attachment from Infancy to Adulthood*, ed. Klaus Grossman, Karin Grossman, and Everett Waters. New York: Guilford.

Cunningham, Michael R., Anita P. Barbee, and Carolyn L. Pike. 1990. "What Do Women Want? Facialmetric Assessment of Multiple Motives in the Perception of Male Facial Physical Attractiveness." *Journal of Personality and Social Psychology* 59: 61–72.

Daily, Catherine M., S. Trevis Certo, and Dan R. Dalton. 1999. "A Decade of Corporate Women: Some Progress in the Boardroom, None in the Executive Suite." *Strategic Management Journal* 20: 93–99.

Damasio, Antonio. 1994. *Descartes' Error: Emotion, Reason and the Human Brain.* New York: HarperCollins.

———. 1996. "The Somatic Market Hypothesis and the Possible Functions of the Prefrontal Cortex." *Philosophical Transactions of the Royal Society, B: Biological Sciences* 351: 1413–20.

Darwin, Charles. 1856. Letter to J. D. Hooker. July 13. Darwin Correspondence Project, http://darwinproject.ac.uk/darwinletters/calendar/entry-1924.html#back-mark-1924.f2, accessed June 19, 2011.

———. [1859] 1979. *The Origin of Species.* New York: Gramercy Books.

———. [1871] 1981. *The Descent of Man, and Selection in Relation to Sex.* Introduction by J. T. Bonner and R. B. May. Princeton, NJ: Princeton University Press.

———. [1872] 1998. *The Expression of the Emotions in Man and Animals.* Introduction, afterword, and commentaries by Paul Ekman. Oxford: Oxford University Press.

Deary, Ian J., Paul Irwing, Geoff Der, and Timothy Bates. 2007. "Brother-Sister Differences in the *g* Factor in Intelligence: Analysis of Full, Opposite-Sex Siblings from the NLSY 1979." *Intelligence* 35: 451–56.

Della Giusta, Marina, Sarah Louise Jewell, and Uma S. Kambhampati. 2011. "Gender and Life Satisfaction in the UK." *Feminist Economics* 17: 1–34.

Desmond, Adrian, and James Moore. 2009. *Darwin's Sacred Cause: Race, Slavery and the Quest for Human Origins.* London: Allen Lane.

Dessalles, Jean-Louis. 1999. "Coalition Factors in the Evolution of Non-kin Altruism." *Advances in Complex Systems* 2: 143–72.

DeSteno, David, Piercarlo Vadesolo, and Monica Y. Bartlett. 2006. "Jealousy and the Threatened Self: Getting to the Heart of the Green-Eyed Monster." *Journal of Personality and Social Psychology* 91: 626–41.

Devlin, B., Michael Daniels, and Kathryn Roeder. 1997. "The Heritability of IQ." *Nature* 388: 468–71.

de Waal, Frans B. M. 1982. *Chimpanzee Politics: Power and Sex among Apes*. New York: Harper and Row.

———. 1989. *Peacemaking among Primates*. Cambridge, MA: Harvard University Press.

———. 2001. *Tree of Origin: What Primate Behavior Can Tell Us about Human Social Evolution*. Cambridge, MA: Harvard University Press.

Diamond, Jared. 1987. "The Worst Mistake in the History of the Human Race." *Discover*. May. www.ditext.com/diamond/mistake.html.

Dohmen, Thomas, Armin Falk, David Huffman, Uwe Sunde, Jürgen Schupp, and Gert G. Wagner. 2011. "Individual Risk Attitudes: Measurement, Determinants, and Behavioral Consequences." *Journal of the European Economic Association* 9: 522–50.

Doniger, Wendy. 1999. *Splitting the Difference: Gender and Myth in Ancient Greece and India*. Chicago: University of Chicago Press.

———. 2000. *The Bedtrick: Tales of Sex and Masquerade*. Chicago: University of Chicago Press.

———. 2005. *The Woman Who Pretended to Be Who She Was*. New York: Oxford University Press.

Dougherty, C. 2005. "Why Are the Returns to Schooling Higher for Women Than for Men?" *Journal of Human Resources* 40(4): 969–88.

"Down or Out: A Black Male Professor Kicks Up a Storm about Black Women and Marriage." 2011. *Economist*. October 5. www.economist.com/node/21532296.

Dreber, A., and M. Hoffman. 2010. "Biological Basis for Sex Differences in Risk Aversion and Competitiveness." Unpublished manuscript.

Dreber, A., and M. Johannesson. 2008. "Gender Differences in Deception." *Economics Letters* 99: 197–99.

Dreber, A., E. von Essen, and E. Ranehill. 2011. "Outrunning the Gender Gap: Boys and Girls Compete Equally." *Experimental Economics* 14: 567–82.

Dreber, A., C. L. Apicella, D. T. A. Eisenberg, J. R. Garcia, R. Zamore, J. K. Lum, and B. C. Campbell. 2009. "The 7R Polymorphism in the Dopamine Receptor D4 gene (DRD4) Is Associated with Financial Risk-Taking in Men." *Evolution and Human Behavior* 30(2): 85–92.

Duckworth, Angela, and Martin Seligman. 2005. "Self-Discipline Outdoes IQ in Predicting Academic Performance of Adolescents." *Psychological Science* 16: 939–44.

Dunbar, Robin. 1992. "Neocortex Size as a Constraint on Group Size in Primates." *Journal of Human Evolution* 20: 469–93.

Edlund, Lena, and Evelyn Korn. 2002. "A Theory of Prostitution." *Journal of Political Economy* 110(1): 181–214.

Ellison, Glenn, and Ashley Swanson. 2010. "The Gender Gap in Secondary School

Mathematics at High Achievement Levels: Evidence from the American Mathematics Competitions." *Journal of Economic Perspectives* 24: 109–28.

Ellison, P. T., and P. B. Gray. 2009. *Endocrinology of Social Relationships.* Cambridge, MA: Harvard University Press.

Empson, William. 1984. *Collected Poems.* London: Hogarth Press.

Engel, C. 2011. "Dictator Games: A Meta Study." *Experimental Economics* 14: 583–610.

Falk, Armin, Ingo Menrath, Johannes Siegrist, and Pablo Emilio Verde. 2011. *Cardiovascular Consequences of Unfair Pay.* Discussion Paper 8463, Centre for Economic Policy Research, London.

Falkinger, Josef. 2007. "Attention Economies." *Journal of Economic Theory* 133: 266–94.

———. 2008. "Limited Attention as the Scarce Resource in Information-Rich Economies." *Economic Journal* 118: 1596–1620.

Farrell, Warren. 1993. *The Myth of Male Power: Why Men Are the Disposable Sex.* New York: Simon and Schuster.

Fernández, Raquel, and Joyce Cheng Wong. 2011. "The Disappearing Gender Gap: The Impact of Divorce, Wages and Preferences on Education Choices and Women's Work." Discussion Paper 8627, Centre for Economic Policy Research, London.

Fessler, D. M. T., and K. J. Haley. 2003. "The Strategy of Affect: Emotions in Human Cooperation." In *Genetic and Cultural Evolution of Cooperation*, ed. Peter Hammerstein. Cambridge, MA: MIT Press.

Fine, Cordelia. 2010. *Delusions of Gender: How Our Minds, Society and Neurosexism Create Difference.* New York: W. W. Norton.

Fischer, C. S. 1994. *America Calling: A Social History of the Telephone to 1940.* Berkeley: University of California Press.

Fisher, Helen. 2004. *Why We Love: The Nature and Chemistry of Romantic Love.* New York: Henry Holt.

Flynn, J. R. 2007. *What Is Intelligence?* Cambridge: Cambridge University Press.

Foreman, Jonathan. 2009. "Taking the Private Jet to Copenhagen." *Sunday Times.* November 29. http://women.timesonline.co.uk/tol/life_and_style/women/celebrity/article6931572.ece.

Forret, Monica L., and Thomas W. Dougherty. 2004. "Networking Behaviors and Career Outcomes: Differences for Men and Women?" *Journal of Organizational Behavior* 25: 419–37.

Foucault, Michel. 1990. *A History of Sexuality.* New York: Vintage.

Fox, Stephen. 1984. *The Mirror Makers: A History of American Advertising and Its Creators.* Chicago: University of Illinois Press.

Frank, Robert. 1988. *Passions within Reason.* New York: W. W. Norton.

Frazzetto, Giovanni. 2010. "The Science of Online Dating." *EMBO Reports* 11: 25–27. doi:10.1038/embor.2009.264.

Freeman, Derek. 1983. *Margaret Mead and Samoa: The Making and Unmaking of an Anthropological Myth.* Cambridge, MA: Harvard University Press.

Friebel, Guido, and Paul Seabright. 2011. "Do Women Have Longer Conversations? Telephone Evidence of Gendered Communication Strategies." *Journal of Economic Psychology* 32: 348–56. doi:10.1016/j.joep.2010.12.008.

Fryer, Roland, and Steven Levitt. 2006. "Testing for Racial Differences in the Mental Ability of Young Children." NBER Working Paper 12066, National Bureau of Economic Research, Cambridge, MA.

Galef, Bennett G. 2008. "Social Influences on the Mate Choices of Male and Female Japanese Quail." *Comparative Cognition and Behavior Reviews* 3: 1–12.

Gallagher, A. M., and J. C. Kaufman, eds. *Gender Differences in Mathematics: An Integrative Psychological Approach.* Cambridge: Cambridge University Press.

Gambetta, Diego. 2009. *Codes of the Underworld: How Criminals Communicate.* Princeton, NJ: Princeton University Press.

Garcia, Guy. 1998. *The Decline of Men.* New York: HarperCollins.

Geary, David C., Scott J. Saults, Fan Lui, and Mary K. Hoard. 2000. "Sex Differences in Spatial Cognition, Computational Fluency, and Arithmetical Reasoning." *Journal of Experimental Child Psychology* 77: 337–53.

Gee, Henry, Rory Howlett, and Philip Campbell. 2009. "Fifteen Evolutionary Gems." *Nature.* January. doi:10.1038/nature07740.

Geletkanycz, Marta, Brian Boyd, and Sydney Finkelstein. 2001. "The Strategic Value of CEO External Directorate Networks: Implications for CEO Compensation." *Strategic Management Journal* 22(9): 889–98.

Gesquière, Laurence R., Niki H. Learn, M. Carolina M. Simao, Patrick O. Onyango, Susan C. Alberts, and Jeanne Altmann. 2011. "Life at the Top: Rank and Stress in Wild Baboons." *Science* 333: 357–60.

Gibson, Robert, and Jacob Höglund. 1992. "Copying and Sexual Selection." *Trends in Ecology and Evolution* 7: 229–32.

Giraudoux, Jean. 1997. *Suzanne et le Pacifique.* Paris: Grasset.

Giridharadas, Anand. 2007. "The Ink Fades on a Profession as India Modernizes." *New York Times.* December 26.

Gneezy, U., M. Niederle, and A. Rustichini. 2003. "Performance in Competitive Environments: Gender Differences." *Quarterly Journal of Economics* 118: 1049–74.

Gneezy, Uri, and Aldo Rustichini. 2004. "Gender and Competition at a Young Age." *American Economic Review* 94(2): 377–81.

Gneezy, U., K. L. Leonard, and J. A. List. 2009. "Gender Differences in Competition: Evidence from a Matrilineal and a Patriarchal Society." *Econometrica* 77(3): 909–31.

Goffman, Erving. 1963. *Stigma: Notes on the Management of Spoiled Identity.* Englewood Cliffs, NJ: Prentice-Hall.

Goldacre, Ben. 2009. *Bad Science.* London: HarperCollins.

Goldin, Claudia, and Lawrence F. Katz. 2002. "The Power of the Pill: Oral Contraceptives and Women's Career and Marriage Decisions." *Journal of Political Economy* 110(4): 730–70.

Goodall, Jane. 1986. "Social Rejection, Exclusion, and Shunning among the Gombe Chimpanzees." *Ethology and Sociobiology* 7: 227–39.

Gouchie, Catherine, and Doreen Kimura. 1991. "The Relationship between Testosterone Levels and Cognitive Ability Patterns." *Psychoneuroendocrinology* 16: 323–34.

Grafen, Alan. 1990. "Biological Signals as Handicaps." *Journal of Theoretical Biology* 144: 517–46.

Granovetter, Mark. 1973. "The Strength of Weak Ties." *American Journal of Sociology* 78: 1360–80.

Green, Eileen, and Carrie Singleton. 2009. "Mobile Connections: An Exploration of the Place of Mobile Phones in Friendship Relations." *Sociological Review* 57: 125–44.

Greenwood, Jeremy, Ananth Seshadri, and Mehmet Yorukoglu. 2005. "Engines of Liberation." *Review of Economic Studies* 72: 109–33.

Gregoric, Aleksandra, Lars Oxelheim, Trond Randoy, and Steen Thomsen. 2010. "How Diverse Can you Get? Gender Quotas and the Diversity of Nordic Boards." Working Paper, Lund Institute of Economic Research, Lund.

GSMA/A. T. Kearney. 2011a. "Africa Mobile Observatory: Driving Economic and Social Development through Mobile Services." www.gsmworld.com/documents/African_Mobile_Observatory_Full_Report_2011.pdf.

———. 2011b. "Asia Pacific Mobile Observatory: Driving Economic and Social Development through Mobile Broadband." www.gsmworld.com/documents/AP%20Mobile%20Observatory%20Full%20Report.pdf.

Günther, Christine, Neslihan Arslan Ekinci, Christiane Schwieren, and Martin Strobel. 2010. "Women Can't Jump? An Experiment on Competitive Attitudes and Stereotype Threat." *Journal of Economic Behavior and Organization* 75: 395–401.

Gupta, Nabanita Datta, Andres Poulsen, and Marie-Claire Villeval. 2005. "Male and Female Competitive Behavior: Experimental Evidence." Working Paper 1833, Institute for the Study of Labor, Bonn.

Gurven, Michael, Hillard Kaplan, and Alfredo Zelada Supa. 2007. "Mortality Experience of Tsimane Amerindians of Bolivia: Regional Variation and Temporal Trends." *American Journal of Human Biology* 19: 376–98.

Haidt, Jonathan. 2007. "The New Synthesis in Moral Psychology." *Science* 316: 998–1002. doi:10.1126/science.1137651.

Halpern, D. F., C. P. Benbow, D. C. Geary, R. C. Gur, J. S. Hyde, and M. A. Gernsbacher. 2007. "The Science of Sex Differences in Science and Mathematics." *Psychological Science in the Public Interest* 8: 1–51.

Hamermesh, D. S., and J. E. Biddle. 1994. "Beauty and the Labor Market." *American Economic Review* 84: 1174–94.

Harcourt, A. H., P. H. Harvey, S. G. Larson, and R. V. Short. 1981. "Testis Weight, Body Weight and Breeding System in Primates." *Nature* 293: 55–57.

Harris, Adam, and Ulrike Hahn. 2011. "Unrealistic Optimism about Future Life Events: A Cautionary Note." *Psychological Review* 118: 135–54.

Hawkes, Kristen. 1991. "Showing Off: Tests of an Hypothesis about Men's Foraging Goals." *Ethology and Sociobiology* 12: 29–54.

———. 2004. "Mating, Parenting and the Evolution of Human Pair Bonds." In *Kinship and Behavior in Primates*, ed. Bernard Chapais and Carol M. Berman. New York: Oxford University Press.

Hawkes, Kristen, James F. O'Connell, and Nicholas Blurton Jones. 1991. "Hunting Income Patterns among the Hadza: Big Game; Common Goods; Foraging Goals and the Evolution of the Human Diet." *Philosophical Transactions of the Royal Society B: Biological Sciences* 334: 243–51.

Hawkes, Kristen, James F. O'Connell, Nicholas Blurton Jones, Helen Alvarez, and Eric L. Charnov. 2000. "The Grandmother Hypothesis and Human Evolution." In *Adaptation and Human Behavior: An Anthropological Perspective*, ed. L. Cronk, N. Chagnon, and W. Irons. New York: Aldine de Gruyter.

Hawkins, Jeff. 2004. *Intelligence*. New York: Henry Holt.

Hedges, Larry V., and Amy Nowell. 1995. "Sex Differences in Mental Test Scores: Variability and Numbers of High Scoring Individuals." *Science* 269: 41–45.

Henazi, S. P., and L. Barrett. 1999. "The Value of Grooming to Female Primates." *Primates* 40: 47–59.

Herrnstein, Richard, and Charles Murray. 1994. *The Bell Curve: Intelligence and Class Structure in American Life*. New York: Free Press.

Hill, Kim, and Magdalena Hurtado. 1996. *Ache Life History: The Ecology and Demography of a Foraging People*. Hawthorne, NY: Aldine de Gruyter.

Hill, Kim, Magdalena Hurtado, and R. S. Walker. 2007. "High Adult Mortality among Hiwi Hunter-Gatherers: Implications for Human Evolution." *Journal of Human Evolution* 52: 443–54.

Hines, Melissa. 2011. "Gender Development and the Human Brain." *Annual Review of Neuroscience* 34: 69–88.

Hobbes, Thomas. 2008. *De Cive*. Kindle edition.

Hoffman, Moshe, Uri Gneezy, and John A. List. 2011. "Nurture Affects Gender Differences in Spatial Abilities." *Proceedings of the National Academy of Sciences* 108: 14786–88.

Hooper, Paul L. 2011. "The Structure of Energy Production and Redistribution among Tsimane' Forager-Horticulturalists." PhD diss., University of New Mexico, Albuquerque.

Hrdy, Sarah Blaffer. 2009. *Mothers and Others: The Evolutionary Origins of Mutual Understanding.* Cambridge, MA: Belknap Press.

Hunt, Earl. 2010. *Human Intelligence.* Cambridge: Cambridge University Press.

Hupka, R. B., and J. M. Ryan. 1990. "The Cultural Contribution to Jealousy: Cross-Cultural Aggression in Sexual Jealousy Situations." *Behavior Science Research* 24: 51–71.

Hyde, Janet Shibley, and Marcia Linn. 1988. "Gender Differences in Verbal Ability: A Meta-Analysis." *Psychological Bulletin* 104: 53–69.

Hyde, Janet S., and Janet E. Mertz. 2009. "Gender, Culture and Mathematics Performance." *Proceedings of the National Academy of Sciences* 106: 8801–7.

Ibarra, Herminia. 1997. "Paving an Alternative Route: Gender Differences in Managerial Networks." *Social Psychology Quarterly* 60: 91–102.

Igarashi, Tasuku, Jiro Takai, and Toshikazu Yoshida. 2005. "Gender Differences in Social Network Development via Mobile Phone Text Messages: A Longitudinal Study." *Journal of Social and Personal Relationships* 22: 691.

International Labor Organization. 2011. *Maternity Protection Database.* www.ilo.org/travaildatabase/servlet/maternityprotection.

International Telecommunication Union. 2011a. *Fixed Internet Subscriptions.* http://itu.int/ITU-D/ict/statistics/index.html.

———. 2011b. *Key ICT Indicators for Developed and Developing Countries and the World (Totals and Penetration Rates).* http://itu.int/ITU-D/ict/statistics/at_glance/KeyTelecom.html.

Irwin, Robert. 2001. "Is This the Man Who Inspired Bin Laden?" *Guardian.* November 1. www.guardian.co.uk/world/2001/nov/01/afghanistan.terrorism3.

Jackson, Douglas N., and J. Philippe Rushton. 2006. "Males Have Greater *g*: Sex Differences in General Mental Ability from 100,000 17- to 18-Year-Olds on the Scholastic Assessment Test." *Intelligence* 34: 479–86.

Jeanquart-Barone, Sandy. 1993. "Trust Differences between Supervisors and Subordinates: Examining the Role of Race and Gender." *Sex Roles* 29: 1–11.

Jeffreys, Sheila. 1997. *The Idea of Prostitution.* Melbourne: Spinifex Press.

Johnson, W., A. Carothers, and I. J. Deary. 2008. "Sex Differences in Variability in General Intelligence." *Perspectives on Psychological Science* 3(6): 518–31.

———. 2009. "A Role for the X Chromosome in Sex Differences in Variability in General Intelligence." *Perspectives on Psychological Science* 4(6): 598–611.

Jorm, Anthony F., Kaarin J. Anstey, Helen Christensen, and Bryan Rodgers. 2004. "Gender Difference in Cognitive Abilities: The Mediating Role of Health States and Health Habits." *Intelligence* 32: 7–23.

Judson, Olivia. 2002. *Dr. Tatiana's Sex Advice to All Creation.* New York: Henry Hudson.

Kaplan, Hillard, Kim Hill, Jane Lancaster, and A. Magdalena Hurtado. 2000. "A Theory of Human Life History Evolution: Diet, Intelligence, and Longevity." *Evolutionary Anthropology* 9: 156–85.

Kaplan, Hillard, Kim Hill, A. Magdalena Hurtado, and Jane Lancaster. 2001. "The Embodied Capital Theory of Human Evolution." In *Reproductive Ecology and Human Evolution*, ed. P. T. Ellison, 293–317. Hawthorne, NY: Aldine de Gruyter.

Kaplan, Hillard, Paul Hooper, and Michael Gurven. 2009. "The Evolutionary and Ecological Roots of Human Social Organization." *Philosophical Transactions of the Royal Society B: Biological Sciences* 364: 3289–99.

Keller, Joseph. 1928. *Belle de Jour.* Paris: Gallimard.

Klingberg, Torkel. 2009. *The Overflowing Brain: Information Overload and the Limits of Working Memory.* Oxford: Oxford University Press.

Knight, Jonathan. 2002. "Sexual Stereotypes." *Nature* 415: 254–56.

Kocoska-Maras L., N. Zethraeus, A. Flöter Rådestad, T. Ellingsen, B. von Schoultz, M. Johannesson, and A. Lindén Hirschberg. 2011. "A Randomized Trial of the Effect of Testosterone and Estrogen on Verbal Fluency, Verbal Memory, and Spatial Ability in Healthy Postmenopausal Women." *Fertility and Sterility* 95: 152–57.

Komisaruk, Barry R., Carlos Beyer-Flores, and Beverly Whipple. 2006. *The Science of Orgasm.* Baltimore, MD: Johns Hopkins University Press.

Kosfeld, Michael, Markus Heinrichs, Paul J. Zak, Urs Fischbacher, and Ernst Fehr. 2005. "Oxytocin Increases Trust in Humans." *Nature* 435: 673–76.

Kostyniuk, Lidia P., Lisa J. Molnar, and David W. Eby. 1996. "Are Women Taking More Risks While Driving? A Look at Michigan Drivers." Proceedings from the Second National Conference on Women's Travel Issues, Baltimore, MD.

Kross, Ethan, Marc Berman, Walter Mischel, Edward Smith, and Tor Wager. 2011. "Social Rejection Shares Somatosensory Representations with Physical Pain." *Proceedings of the National Academy of Sciences* 108: 6270–75.

Kuhn, Peter, and Marie-Claire Villeval. 2011. "Do Women Prefer a Cooperative Work Environment?" Unpublished manuscript.

Kulich, Clara, Grzegorz Troianowski, Michelle K. Ryan, S. Alexander Haslam, and Luc Renneboog. 2011. "Who Gets the Carrot and Who Gets the Stick? Evidence of Gender Disparities in Executive Remuneration." *Strategic Management Journal* 30: 301–21.

Kunda, Ziva. 1990. "The Case for Motivated Reasoning." *Psychological Bulletin* 108: 480–98.

Lalanne, Marie, and Paul Seabright. 2011. "The Old Boy Network: Gender Difference in the Impact of Social Networks on Remuneration in Top Executive Jobs." Discussion Paper 8623, Centre for Economic Policy Research, London.

Ledford, Heidi. 2008. " 'Monogamous' Vole in Love-Rat Shock." *Nature* 451: 617.

Leduc, Claudine. 1992. "Marriage in Ancient Greece." In *A History of Women: From*

Ancient Goddesses to Christian Saints, ed. Pauline Schmitt Pantel. Cambridge, MA: Harvard University Press.

Lemish, Dafna, and Akiba A. Cohen. 2005. "On the Gendered Nature of Mobile Phone Culture in Israel." *Sex Roles* 52: 7–8.

Levitt, Steven D., and Sudhir Alladi Venkatesh. 2007. "An Empirical Analysis of Street-Level Prostitution." Unpublished manuscript.

Lindquist, Kristen A., Tor D. Wager, Hedy Kober, Eliza Bliss-Moreau, and Lisa Feldman Barrett. 2011. "The Brain Basis of Emotion: A Meta-analytic Review." *Behavioral and Brain Sciences*, in press.

Lloyd, Elisabeth A. 2005. *The Case of the Female Orgasm: Bias in the Science of Evolution*. Cambridge, MA: Harvard University Press.

Low, Bobbi S. 2000. *Why Sex Matters: A Darwinian Look at Human Behaviour*. Princeton, NJ: Princeton University Press.

Lynn, Richard. 1999. "Sex Differences in Intelligence and Brain Size: A Developmental Theory." *Intelligence* 27: 1–12.

Lynn, Richard, and Paul Irwing. 2004. "Sex Differences on the Progressive Matrices: A Meta-analysis." *Intelligence* 32: 481–98.

MacDonald, Ian, Bethany Kempster, Liana Zanette, and Scott A. Macdougall-Shackleton. 2006. "Early Nutritional Stress Impairs Development of a Song-Control Brain Region in Both Male and Female Song-Sparrows *Melospiza melodia* at the Onset of Song Learning." *Proceedings of the Royal Society B* 273: 2559–64.

Manning, Alan, and Farzad Saidi. 2010. "Understanding the Gender Pay Gap: What's Competition Got to Do with It?" *Industrial and Labor Relations Review* 63(4): 681–98.

Manning, Alan, and Joanna Swaffield. 2008. "The Gender Gap in Early-Career Wage Growth." *Economic Journal* 118: 983–1024.

Marlowe, Frank. 2010. *The Hadza: Hunter-Gatherers of Tanzania*. Berkeley: University of California Press.

Marmot, Michael. 2004. *The Status Syndrome: How Social Standing Affects Our Heath and Longevity*. New York: Times Books.

Mather, Mark, and Dia Adams. 2007. "The Crossover in Female-Male College Enrollment Rates." Population Reference Bureau. www.prb.org/Articles/2007/CrossoverinFemaleMaleCollegeEnrollmentRates.aspx.

Mayhew, Daniel R., Susan A. Ferguson, Katharine J. Desmond, and Herbert M. Simpson. 2003. "Trends in Fatal Crashes Involving Female Drivers, 1975–1998." *Accident Analysis and Prevention* 35: 407–15.

Mayhew, Henry. [1861] 1968. *London Labour and the London Poor*. Facsimile ed. New York: Dover Publications.

Mead, Margaret. [1928] 1973. *Coming of Age in Samoa*. New York: American Museum of Natural History.

Meston, Cindy, and David Buss. 2009. *Why Women Have Sex*. New York: Random House.

Mikulincer, Mario, and Gail S. Goodman, eds. 2006. *Dynamics of Romantic Love: Attachment, Cargiving and Sex*. New York: Guilford Press.

Milinski, Manfred. 2003. "Perfumes." In *Evolutionary Aesthetics*, ed. Eckart Voland and Karl Grammer, 325–39. Berlin: Springer.

Miller, Geoffrey. 2000. *The Mating Mind: How Sexual Choices Shaped the Evolution of Human Nature*. New York: Anchor.

———. 2009. *Spent: Sex, Evolution, and Consumer Behavior*. New York: Viking Penguin.

Milner, B., S. Corkin, and H. L. Teuber. 1968. "Further Analysis of the Hippocampal Amnesic Syndrome: 14-Year Follow-Up Study of H.M." *Neuropsychologia* 6: 215–34.

Mirrlees, James. 1997. "Information and Incentives: The Economics of Carrots and Sticks." *Economic Journal* 107: 1311–29.

Mobius, M. M., and T. S. Rosenblat. 2006. "Why Beauty Matters." *American Economic Review* 96: 222–35.

Moore, Gwen. 1990. "Structural Determinants of Men's and Women's Personal Networks." *American Sociological Review* (55): 726–35.

Müller, Gerrit, and Erik Plug. 2006. "Estimating the Effect of Personality on Male-Female Earnings." *Industrial and Labor Relations Review* 60: 3–22.

Muller, Martin N., and Richard W. Wrangham. 2009. *Sexual Coercion in Primates and Humans: An Evolutionary Perspective on Male Aggression against Females*. Cambridge, MA: Harvard University Press.

Nasscom. 2009. *Nasscom IT Industry Factsheet*. www.nasscom.in, accessed June 29, 2011.

Niederle, M., and L. Vesterlund. 2007. "Do Women Shy Away from Competition? Do Men Compete Too Much?" *Quarterly Journal of Economics* 122: 1067–101.

———. 2008. "Gender Differences in Competition." *Negotiation Journal* 24(4): 447–63.

Nishida, Toshisada. 1996. "Coalition Strategies among Adult Male Chimpanzees of the Mahale Mountains, Tanzania." In *Great Ape Societies*, ed. William C. McGrew, Linda F. Marchant, and Toshisada Noshida. Cambridge: Cambridge University Press.

Nyhus, Ellen K., and Empar Pons. 2005. "The Effects of Personality on Earnings." *Journal of Economic Psychology* 26: 363–84.

OECD. 2011a. *Labour Force Statistics (Harmonised Unemployment Rates)*. Organisation for Economic Co-operation and Development. http://stats.oecd.org.

——. 2011b. *Education and Training Statistics: New Entrants by Sex and Age*. Organisation for Economic Co-operation and Development. http://stats.oecd.org.

Ogilvie, Sheilagh. 2003. *A Bitter Living: Women, Markets, and Social Capital in Early Modern Germany*. Oxford: Oxford University Press.

——. 2011. *Institutions and European Trade: Merchant Guilds, 1000–1800*. Cambridge: Cambridge University Press.

Owren, M.-J., and J.-A. Bachorowski. 2001. "The Evolution of Emotional Expression: A 'Selfish-Gene' Account of Smiling and Laughter in Early Hominids and Humans." In *Emotions: Current Issues and Future Directions*, ed. T. J. Mayne and G. A. Bonnano. New York: Guilford Press.

Packer, C., and A. E. Pusey. 1983. "Adaptations of Female Lions to Infanticide by Incoming Males." *American Naturalist* 121: 716–28.

Pandit, S. A., and C. P. van Schaik. 2003. "A Model for Leveling Coalitions among Primate Males: Toward a Theory of Egalitarianism." *Behavioral Ecology and Sociobiology* 55: 161–68.

Pantel, P. S., ed. 1992. *A History of Women: From Ancient Goddesses to Christian Saints*. Vol. 1. Cambridge, MA: Belknap Press.

Parker, Kathleen. 2008. *Save the Males: Why Men Matter, Why Women Should Care*. New York: Random House.

Pessoa, Fernando. 1987. *Le Gardeur de troupeaux, et les autres poèmes d'Alberto Caeiro avec Poésies d'Alvaro de Campos,* trans. Armand Guibert. Paris: Gallimard.

Pinker, Steven. 2011. *The Better Angels of Our Nature: The Decline of Violence in History and Its Causes*. New York: Allen Lane.

Posner, Richard A. 1994. *Sex and Reason*. Cambridge, MA: Harvard University Press.

Power, Margaret. 2005. *The Egalitarians: Human and Chimpanzee; An Anthropological View of Social Organization*. New York: Cambridge University Press.

Qutb, Sayyid. 1964. *Milestones*. Cairo: Kazi.

Ramanujan, A. K., ed. and trans. 1985. *Poems of Love and War: From the Eight Anthologies and the Ten Long Poems of Classical Tamil*. New York: Columbia University Press.

Ramsden, Sue, Fiona M. Richardson, Goulven Josse, Michael S. C. Thomas, Caroline Ellis, Clare Shakeshaft, Mohamed L. Seghier, and Cathy J. Price. 2011. "Verbal and Non-verbal Intelligence Changes in the Teenage Brain." *Nature* 479: 113–16.

Reid, J. M., P. Arcese, A. L. E. V. Cassidy, S. M. Hiebert, J. N. M. Smith, P. K. Stoddard, A. B. Marr, and L. F. Keller. 2005. "Fitness Correlates of Song Repertoire Size in Free-Living Song Sparrows (*Melospizia melodia*)." *American Naturalist* 165: 299–310.

Ridley, Matt. 1993. *The Red Queen: Sex and the Evolution of Human Nature*. New York: Perennial, HarperCollins.

———. 2004. *Nature via Nurture*. New York: HarperCollins.

Ringdal, Nils Johan. 1997. *Love for Sale: A World History of Prostitution*. New York: Grove.

Rosin, Hanna. 2010. "The End of Men." *Atlantic*. July–August.

Roughgarden, Joan. 2004. *Evolution's Rainbow: Diversity, Gender, and Sexuality in Nature and People*. Berkeley: University of California Press.

———. 2009. *The Genial Gene: Deconstructing Darwinian Selfishness*, Berkeley: University of California Press.

Rushton, J. P., and A. R. Jensen. 2005. "Thirty Years of Research on Race Differences in Cognitive Ability." *Psychology, Public Policy and Law* 11: 235–94.

Ryan, Christopher, and Cacilda Jetha. 2010. *Sex at Dawn: The Prehistoric Origins of Modern Sexuality*. New York: Harper.

Ryan, Michelle K., and S. Alexander Haslam. 2005. "The Glass Cliff: Evidence that Women Are Over-represented in Precarious Leadership Positions." *British Journal of Management* 16: 81–90.

Salgado, J. F. 1997. "The Five-Factor Model of Personality and Job Performance in the European Community." *Journal of Applied Psychology* 82: 30–43.

Sapolsky, Robert. 2005. *Monkeyluv*. New York: Scribner.

Säve-Söderbergh, Jenny. 2011. *Are Women Asking for Low Wages? Gender Differences in Competitive Bargaining Strategies and Ensuing Bargaining Success*. Working Paper 2007:07, Swedish Institute for Social Research, Stockholm.

Scharlemann, J. P. W., C. C. Eckel, A. Kacelnik, and R. K. Wilson. 2001. "The Value of a Smile: Game Theory with a Human Face." *Journal of Economic Psychology* 22: 617.

Schmitt, David P., Anu Realo, Martin Voracek, and Jüri Allik. 2008. "Why Can't a Woman Be More Like a Man? Sex Differences in Big Five Personality Traits across 55 Cultures." *Journal of Personality and Social Psychology* 94: 168–82.

Scotchmer, Suzanne. 2008. "Risk Taking and Gender in Hierarchies." *Theoretical Economics* 3: 499–524.

Scott, Dow. 1983. "Trust Differences between Men and Women in Superior-Subordinate Relationships." *Group and Organization Studies* 8: 319–36.

Seabright, Paul. 2009. "Continuous Preferences and Discontinuous Choices: How Altruists Respond to Incentives." *B.E. Journal of Theoretical Economics* 9. doi:10.2202/1935-1704.1346.

———. 2010. *The Company of Strangers: A Natural History of Economic Life*. Rev. ed. Princeton, NJ: Princeton University Press.

Searcy, William A., and Stephen Nowicki. 2008. "Bird Song and the Problem of Honest Communication." *American Scientist* 96: 114–21.

Sexton, Steven E., and Alison L. Sexton. 2011. "Conspicuous Conservation: The Prius

Effect and the Willingness-to-Pay for Environmental Bona Fides." Unpublished manuscript.

Shih, M., T. L. Pittinsky, and N. Ambady. 1999. "Stereotype Susceptibility: Identity, Salience and Shifts in Quantitative Performance." *Psychological Science* 10: 80–83.

Silk, J. B. 2003. "Cooperation without Counting: The Puzzle of Friendship." In *Genetic and Cultural Evolution of Cooperation*, ed. Peter Hammerstein. Cambridge, MA: MIT Press.

Silk, J. B., S. Alberts, and J. Altmann. 2004. "Patterns of Coalition Formation by Adult Female Baboons in Amboseli, Kenya." *Animal Behavior* 67: 573–82.

Simmons, Leigh W. 2001. *Sperm Competition and Its Evolutionary Consequences in the Insects.* Princeton, NJ: Princeton University Press.

Small, Deborah, Linda Babcock, Michele Gelfand, and Hilary Gettman. 2007. "Who Goes to the Bargaining Table? The Influence of Gender and Framing on the Initiation of Negotiation." *Journal of Personality and Social Psychology* 93: 600–613.

Smoreda, Zbigniew, and Christian Licoppe. 2000. "Gender-Specific Use of the Domestic Telephone." *Social Psychology Quarterly* 63: 238–52.

Smuts, Barbara. 1999. *Sex and Friendship in Baboons.* 2nd ed. Cambridge, MA: Harvard University Press.

Soares, Rachel, Jan Combopiano, Allyson Regis, Yelena Shur, and Rosita Wong. 2010. "2010 Catalyst Census: Fortune 500 Women Board Directors." New York: Catalyst, Inc. http://catalyst.org/publication/460/2010-catalyst-census-fortune-500-women-board-directors.

Sommer, Volker, and Paul L. Vasey. 2006. *Homosexual Behavior in Animals: An Evolutionary Perspective.* Cambridge: Cambridge University Press.

Spence, Michael. 1974. *Market Signaling: Informational Transfer in Hiring and Related Screening Processes.* Cambridge, MA: Harvard University Press.

Stanford, Craig B. 1999. *The Hunting Apes: Meat-Eating and the Origins of Human Behavior.* Princeton, NJ: Princeton University Press.

Stearns, P. N. 2000. *Gender in World History.* London: Routledge.

Steckel, Richard, and John Wallis. 2009. "Stones, Bones, Cities and States: A New Approach to the Neolithic Revolution." Unpublished manuscript.

Stendhal. [1830] 1962. *Le rouge et le noir.* Paris: Éditions du Dauphin.

Sterelny, Kim. 2003. *Thought in a Hostile World.* Oxford: Blackwell.

———. 2012. *The Evolved Apprentice.* Cambridge, MA: MIT Press.

Stevenson, Betsey, and Justin Wolfers. 2009. "The Paradox of Declining Female Happiness." *American Economic Journal: Economic Policy* 1: 190–225. doi:10.1257/pol.1.2.190.

Sukumar, R., and M. Gadgil. 1988. "Male-Female Differences in Foraging on Crops by Asian Elephants." *Animal Behaviour* 36: 1233–35.

Sutter, Matthias, Ronald Bosman, Martin G. Kocher, and Frans van Winden. 2009. "Gender Pairing and Bargaining: Beware the Same Sex!" *Experimental Economics* 12: 318–31.

Tannen, Deborah. 1990. *You Just Don't Understand: Women and Men in Conversation.* New York: HarperCollins.

———. 1994. *Talking from 9 to 5: Women and Men at Work; Language, Sex and Power.* Virago Press.

Tanner, Michael, and Bernard Williams. 1976. Comment on E. L. Anscombe, "Contraception and Chastity." In *Ethics and Population*, ed. M. D. Bayles. Cambridge, MA: Schenkman.

Therborn, Göran. 2004. *Between Sex and Power: Family in the World, 1900–2000.* London: Routledge.

Thornhill, Randy, and Steven W. Gangestad. 2008. *The Evolutionary Biology of Human Female Sexuality.* New York: Oxford University Press.

Tiger, Lionel. 1999. *The Decline of Males.* New York: St. Martin's.

Trivers, Robert L. 1972. "Parental Investment and Sexual Selection." In *Sexual Selection and the Descent of Man*, ed. Bernard Campbell. Chicago: Aldine.

———. 2000. "The Elements of a Scientific Theory of Self-Deception." *Annals of the New York Academy of Sciences* 907: 114.

———. 2011. *The Folly of Fools: The Logic of Deceit and Self-Deception in Human Life.* New York: Basic Books.

Tungate, Mark. 2007. *Adland: A Global History of Advertising.* London: Kogan Page.

Uller, Tobias, and L. Christoffer Johansson. 2003. "Human Mate Choice and the Wedding Ring Effect: Are Married Men More Attractive?" *Human Nature* 14: 267–76.

US Census Bureau. 2011. *Educational Attainment in the United States: 2010; Detailed Tables.* www.census.gov/hhes/socdemo/education/data/cps/2010/tables.html.

Valdesolo, Piercarlo, and David DeSteno. 2008. "The Duality of Virtue: Deconstructing the Moral Hypocrite." *Journal of Experimental Social Psychology* 44: 1334–38.

van Schaik, Carel P., and Charles H. Janson, eds. 2000. *Infanticide by Males and Its Implications.* Cambridge: Cambridge University Press.

van Schaik, C. P., S. A. Pandit, and E. R. Vogel. 2004. "A Model for Within-Group Coalitionary Aggression among Males." *Behavioral Ecology and Sociobiology* 57: 101–9.

———. 2005. "Toward a General Model for Male-Male Coalitions in Primate Groups." In *Cooperation in Primates and Humans: Mechanisms and Evolution*, ed. P. M. Kappeler and C. P. van Schaik, 151–71. Heidelberg: Springer.

Veblen, Thorstein 1915: *The Theory of the Leisure Class: An Economic Study of Institutions.* London: Macmillan.

Veyne, Paul. 1996: *Le pain et le cirque: Sociologie historique d'un pluralisme politique.* Paris: Seuil.

Visscher, Peter M., William G. Hill, and Naomi R. Wray. 2008. "Heritability in the Genomics Era: Concepts and Misconceptions." *Nature Reviews: Genetics* 9: 255–66.

von Drehle, David. 2006. "A Lesson in Hate." *Smithsonian.* February. http://smithsonianmag.com/history-archaeology/presence-feb06.html.

Wajman, Judy, Michael Bittman, and Jude Brown. 2009. "Intimate Connections: The Impact of the Mobile Phone on Work/Life Boundaries." In *Mobile Technologies: From Telecommunications to Media*, ed. Gerard Goggin and Larissa Hjorth. London: Routledge.

Wallner, B., and J. Dittami. 1997. "Postestrus Anogenital Swelling in Female Barbary Macaques: The Larger, the Better?" *Annals of the New York Academy of Sciences* 807: 590.

Walum, H., L. Westberg, S. Henningsson, J. M. Neiderhiser, D. Reiss, W. Igl, J. M. Ganiban, et al. 2008. "Genetic Variation in the Vasopressin Receptor 1a Gene (AVPR1A) Associates with Pair-Bonding Behavior in Humans." *Proceedings of the National Academy of Sciences* 105(37): 14153–56.

Wasson, Sam. 2010. *Fifth Avenue, 5 a.m.: Audrey Hepburn, Breakfast at Tiffany's and the Dawn of the Modern Woman.* New York: HarperCollins.

Weatherhead, Patrick J., and Raleigh J. Robinson. 1979. "Offspring Quality and the Polygyny Threshold." *American Naturalist* 113: 201–8.

Whitchurch, E. R., T. D. Wilson, and D. T. Gilbert. 2010. "He Loves Me, He Loves Me Not." *Psychological Science.* December 17.

Wilder, J. A., Z. Mobasher, and M. F. Hammer. 2004. "Genetic Evidence for Unequal Effective Population Sizes of Human Females and Males." *Molecular Biology and Evolution* 21: 2047–57.

World Health Organization. 2011. Global Health Observatory Data Repository. www.who.int.

Wrangham, Richard. 2009. *Catching Fire: How Cooking Made Us Human.* New York: Basic Books.

Wrangham, Richard, and Dale Peterson. 1996. *Demonic Males: Apes and the Origins of Human Violence.* Boston: Mariner.

Wrangham, Richard, James Holland Jones, Greg Laden, David Pilbeam, and Nancy-Lou Conklin-Brittain. 1999. "The Raw and the Stolen." *Current Anthropology* 40: 567–94.

Young, Larry J. 2009. "Love: Neuroscience Reveals It All." *Nature* 457: 148.

Yoshino, Kenji. 2006. *Covering: The Hidden Assault on Our Civil Rights.* New York: Random House.

Zahavi, Amotz. 1975. "Mate Selection: A Selection for a Handicap." *Journal of Theoretical Biology* 53: 205–14.

Zechner, Ulrich, Monika Wilda, Hildegard Kehrer-Sawatzki, Walther Vogel, Rainald Fundele, and Horst Hameister. 2001. "A High Density of X-Linked Genes for General Cognitive Ability: A Run-Away Process Shaping Human Evolution?" *Trends in Genetics* 17: 697–701.

Zelizer, Viviana. 2006. "Money, Power and Sex." *Yale Journal of Law and Feminism* 18: 303–20.

INDEX

ability score, 94
accountancy, 99
actors, 27, 32, 148–149
advertising and advertisements, 27–35, 49, 130, 167, 206
African Americans, 156, 204
after-sales service, 39
aggression, 82, 126, 171
aggressiveness in bargaining, 174
agreeableness, 63, 105–107, 144
agricultural society, 24
  effects on women, 67–68
airline pilots, 100, 112, 177
Allen, Woody, 106
alliances, 6, 126, 183
alpha animals, 64, 159, 162
altruism, 17, 199
American Tobacco Company, 99
anatomy, 38, 68, 71
Anna Karenina (Tolstoy), 59
Antarctic fish, 52
anteaters, 52
anthropology and anthropologists, 34, 72–74, 78, 157, 160, 163, 192, 204
Apollo at Delphi, temple of, 54
Apter, Terri, 175
architects, 100
Arctic fish, 52
arithmetic tests, 94, 197
Arnhem Zoo, 126
asymmetry, 14, 15, 72, 184
athletic tests, 94
attention deficit, 177

Babcock, Linda, 116, 120, 200
bankers, 121, 147
bargaining, aggressiveness in, 174
bargaining power, 64, 72, 75, 76, 78–79, 82, 83, 84
  female, 75–76, 78, 84
basketball and basketball players, 94, 95, 137–138
Beagle (ship), 51
bedbugs, 7, 14, 16, 73
behavior
  predictable, 41
  unpredictable, 184
Belle de Jour (Keller), 59
Belle du Seigneur (Cohen), 3, 27
Bertrand, Marianne, 120–122, 139, 199, 200, 201

beta animals, 159
"Big Five" personality measures, 105–106
black dye, 171
board members, 100, 134–136
Boehm, Christopher, 83
Bonaparte, Marie, 111
bonobos, 6–7, 20, 66–67, 71–72, 78, 81–82, 158, 160
  homosexuality among, 6
borrowing network, 132
Brahmins, 141–142
brain, 7, 14, 28, 40–41, 165–166, 170, 187, 190, 196
  differences between (human) male and female, 22, 71
  of H.M., 105
  large (human), 21, 24, 76–78, 82, 163, 180, 192
  male (human), 17
  of sparrow, 38
  twenty-first-century, 25–26, 87, 153
bread and circuses, 35, 49
Breakfast at Tiffany's (film), 171–172
Brel, Jacques, 56, 157
bribe
  bread and circuses, 35, 49
  sexual, 4, 7
Burt, Ronald, 132
business executives, 95, 137, 158, 176

cannibalism, in praying mantises and spiders, 8
capitalism, 165
car accident, likelihood of, 95
car workers, 147
Carcassonne, 36
care for offspring, 18
  human, 6, 39
career interruptions, 120–121, 179
Case, Anne, 95–96
cell phones (mobile phones), 37, 50, 100, 133, 142, 143, 144
CEOs, 121–122, 128, 137, 176, 179, 195, 199
Charlemagne, 36
charm, 3–4, 25, 32, 43, 47, 152–156
  scarcity of, 141–145
  in the workplace, 48, 146–151
chauffeurs, 100
chefs, 101, 149
chief executives, 121–122, 128, 137, 176, 179, 195, 199
children, 6, 8, 11, 13, 19, 21, 30, 43, 51, 57, 71, 74, 77, 81, 90, 99, 145, 159–160, 184, 191, 195–196, 200
  care of as means of signaling conscientiousness, 124–125
  career breaks and, 120–121, 122, 124, 153, 174, 175, 178–179
  child prostitution, 207
  child rearing, 21, 74, 174
  childhood, 6, 112, 142
  constraint on full-time working, 99, 136, 179, 208–209
  psychometric tests of, 94, 96, 115
chimera, 74
Chimpanzee Politics (de Waal), 64
chimpanzees, 4, 7, 20, 64, 66, 71, 75, 78, 81–82, 126–128, 158–159, 204–205
cigarettes, 8, 99, 167
  advertisements, 167
Cinderella, 152–153
circular reasoning, 95
clams, 95
Clinton, Bill, 162
coalition building, 7, 22, 62–64, 83, 126–129, 139–140, 161–163, 190
coevolution, 52
cognitive abilities, 41, 68–70, 96–98, 101, 105–107, 138
cognitive bias, 41
cognitive capacity, 138
cognitive challenges, 71
cognitive development, 108

cognitive difference, 68–71, 105
cognitive talent, 70
cognitive tests, 68, 96, 101
Cohen, Albert, 3, 27
college degree, 120, 143, 154
collusion and colluding, 81, 118
commercials
  advertising, 31–32, 38
  and sex, 38, 206–207
  television, 31–32
commitment, 12, 21, 42, 45, 56, 57, 122, 129, 138, 139, 155, 164, 176, 178
communication under stress, minimal, 55
comparisons, of multidimensional options, 104
competition, between men and women, 65, 114
computer programmers, 100, 150, 151
computer support technicians, 150–151
conscientiousness, 34–35, 39, 105–107, 125, 147
conspicuous consumption, 35
contacts, personal, 129–130
contextual factors, 101
contraception, 14, 42, 53, 99, 161–165, 195
convergent evolution, 52
conversation, 9, 30, 35, 132, 133–134
cooking, 77
cooks, 101
cooperation, 3–12, 23, 42, 49, 57, 60–67, 72, 75, 76, 82–85, 115, 127, 161, 164, 169, 199, 207
  cooperative breeding, 75, 205
correlation
  between driving fast and accidents, 95
  between height and economic success, 95–96, 110
  between stress and rank, 63
  between talent and height, 95–96, 103
  between test scores (*g*) and economic outcomes, 103
  between test scores (*g*) and height, 96, 103
cortisol, 64
cosmetics, 3, 19, 30, 53
covering (social phenomenon), 30–31, 124, 186
co-workers, 130–131, 201
crisis of men, 23, 90–91, 154–156

Damasio, Antonio, 41, 187
Dame Carcas, 36
dance flies (*Rhamphomyia longicauda*), 4–7, 14, 29–31, 58, 65, 89
Darwin, Charles, 18, 50–53, 60–63, 69, 85, 140, 184, 189, 190, 191
de Waal, Frans, 64, 126, 190, 201, 205
deception
  female gains, 7, 18, 30
  hollow sacs in dance flies, 29
  male gains, 7, 18
  self-deception, 55–56, 189
  silicone breast enhancement, 30
Descartes' Error (Damasio), 41
Descent of Man, The (Darwin), 50, 62, 69
Desmond, Adrian, 51, 189
Diamond, Jared, 85, 193
differential motivation, 22
differential talent, 22
discrimination, 118
Disraeli, Benjamin, 55
divergent evolution, 50–52
divided workplace, 168, 174–178
division of labor, 22, 88, 98, 139, 174
DNA, 18–19, 27–28, 53, 70, 90
dolphins, 7, 127, 158, 183, 191
domestic servants, 164
dominance, 52, 64, 66–67
Donne, John, 56, 57
driving
  fast, 94–95

driving (*continued*)
  on the right, 19–20
  skills, 109
Duchenne, Guillaume, 48
dunnock, 75

earnings differences, 100, 107, 120–122, 136
earnings spread, 149–151, 204
economic outcomes, 24, 94–96, 108, 110, 119, 129
economic power, share of, 95
economists, 36, 42, 46, 95, 106, 130, 156, 188
education
  gender gap in, 91, 107, 154–156
  men's, 90–91, 100, 120, 154–156
  women's, 90–91, 100, 120, 154–156, 176, 194, 204
egg donation, 208, n. 18
eggs, female, 12–13, 15, 29, 39, 84
electricians, 100
emotional stability, 106, 198
emotions, 15, 16, 21, 23–26, 39–45, 56, 68–69, 74, 106, 162, 164, 166, 168, 171, 184, 187, 198
employers negotiating in the workplace, 111–112, 117
employment, formal barriers to, 98
employment discrimination, 118
Empson, William, 93
evolution, 4
  coevolution, 52
  convergent, 52
  of cooperative behavior, 75
  divergent, 50–52
  of homosexuality, 4, 6
  of humans, 6–7, 20, 51, 64, 71, 77, 108, 161, 166, 168–169, 173, 187, 190–191, 192, 205
  of peacock's tail, 49–50
  of sexual behavior, 18–20
exclusion, unjustified, 95–96, 103
executives, 37, 95, 100, 134–136, 158, 176, 177, 201, 203
experience, openness to, 105–106
experiments, 11–12, 22, 46–47, 50, 84, 98, 112–113, 114, 115, 116, 139, 204
Expression of the Emotions in Man and Animals, The (Darwin), 69
externalities, 130, 178
extraversion, 106

fast food chefs, 149
female autonomy, 67, 72, 81, 84
female labor participation, 100, 117, 195
female libido, 17–18
female representation in the workplace
  over-, 117, 124
  under-, 101, 117
female sex cells, 12–14
fetus, 13, 19, 74
fighting, 29, 126, 158, 159
film and video editors, 148, 149
first-class plane travel, 37
fish, 52
fitness, 29, 30, 38, 40, 49, 50, 52, 63, 140, 164, 169, 180
flexible working hours, 136
Flynn effect, 102, 192
foraging, 67, 68, 76, 81, 127, 158, 159
force versus persuasion, as male sexual strategy, 58, 72
Ford, Harrison, 37
Foreman, Jonathan, 36–37, 186
formal barriers to employment, 98
Fortune 500, 100, 115, 121, 128
Frank, Robert, 42, 187
Freud, Sigmund, 111
Friebel, Guido, 132, 202
friendship, 6, 42, 129

g (measure of intelligence), 102–103, 108, 197
  variance in, in men, 107–109, 197, 198

gaffe, defined by Kinsley, 54
Gambetta, Diego, 44, 187
gamete, 21, 84, 184, 188, 208
garter snake, 52
gender gap, 107, 118, 198
gender quota, 199
General Social Survey, 130
genes, 27–28, 30, 39, 52, 107, 108, 110, 188–189, 196
  for lysozyme, 52
  sparrow, 39
genetic drift, 69
genetically determined traits, 107–108, 196
Giraudoux, Jean, 142, 203
Givenchy, Hubert, 171
Gladstone, William, 55
Goffman, Erving, 30
Goldin, Claudia, 120–122, 139, 195, 200, 201
Goodall, Jane, 127, 201
gorillas, 8, 20, 66–67, 70–71, 72, 159
Gowin, E. B., 95, 194
grandmother hypothesis, 192
Granovetter, Mark, 127, 129, 201
Grant, Cary, 34, 35
grooming, 128
grudges, female, 127

"handicap" principle, 49–50
happiness, 22, 89, 141, 156, 169, 193, 204
Harcourt, Alexander, 66, 191
headhunting, 130
health care practitioners, 100
height
  discrimination based on, 118
  increase in, 96
  premium, 96
  representation of people in companies by, 93
  as a signal, 50, 98
Hepburn, Audrey, 171–172
heritable traits, 102, 105
Hierarchy in the Forest (Boehm), 83
high jump, 94, 95
high-status individuals, 161
higher education, 90, 100, 120, 154, 188
H.M. (patient), 105
Hobbes, Thomas, 60, 63, 190
homeopathic medicine, 119
Homo sapiens, 14, 50–51, 61, 68, 82, 83, 128, 158, 159, 160–162, 168, 180
homosexuality, 6, 31, 183, 205
hormonal factors, 116
hormones, 26, 41, 64
  oxytocin, 45
  vasopressin, 45
hours of work, weekly, 120, 121–122
Hrdy, Sarah Blaffer, 74, 81, 192, 193
Hunt, Earl, 103, 191, 197, 198
hunter-gatherers, 21, 22, 41, 67, 68, 71, 72, 74, 76, 78, 79, 82, 83, 84, 89, 118, 125, 162–166
  society, 79, 83, 164
  gender bargain among, 67, 78

income, relative, 22
Index of Movers and Shakers (IMS), 134–136
Industrial Revolution, 166
inequality
  economic, 16, 174
  gender, 174
infectious disease, 89
infidelity, 17, 164, 169, 170, 172–173, 184
  and politicians, 173
instinct, 16, 17, 24
Internet, 144–145, 152–153, 166, 203
IQ, 102, 106–107, 191, 196
irrelevant inclusion, 95–96, 103

jealousy, 8, 19, 74, 163, 171, 206
Jetha, Cacilda, 72, 73
job-finding, 129–130, 145, 202

Kafka, Franz, 94
Katz, Laurence, 120–122, 139, 195, 200, 201
Khan Academy, 144, 203
kin, 28, 130–131
Kinsley, Michael, 54

labor-saving devices, 99, 195
Lalanne, Marie, 134, 202, 208
Laschever, Sara, 116, 120, 200
lawyers, 100
letter-writing, 141–142
Lewinsky, Monica, 162
literacy, 141–142, 144–145
"little black dress," 171–172
Low, Bobbi S., 127, 139
luck, 82, 84, 113–114, 126, 199
Lumière Brothers, 146
lysozyme, 52

Madame de Rênal, 43
makeup artists, 149
male sex cells, 12, 14
manipulation
  of emotions, 25, 41, 45
  in sex, 25
  in signaling and advertisements, 31, 33
  in songs, 32
  and suspicion, 57
Marmot, Michael, 63, 190
Marxists, 118
mathematical skills, 96
Max Planck Institute of Evolutionary Biology, 46
Mayhew, Henry, 142, 198, 203
MBA earnings, 120
Mead, Margaret, 73
merchant guilds, 98
Miller, Geoffrey, 34, 108
minimal communication under stress, 55
mistrust, 9
mitochondria, 90
mobile phones (cell phones), 37, 50, 100, 133, 142–144, 166
model relationship, 168–169, 172–173
monogamy, 17, 19, 73–75
Moore, Gwen, 130
Moore, James, 51
motherhood, 120
motivation, differential, 22
Muller, Martin, 64, 190
multidimensional options, comparisons of, 104
murder, 170
musical recording, 58
mutual signaling, 123

natural selection, 6–8, 15–19, 21, 24, 30, 39, 45, 48, 52–53, 56–59, 68, 70–71, 73–74, 76–77, 82, 90, 145, 163–164, 168, 189, 190, 206
  compatibility with conflict, 7, 53, 59, 73, 169–170
  cooperation and, 57, 62, 85
  emotions and, 16, 40–43, 56, 164
  risk-taking and, 113
negotiation, in the workplace, 111–112, 117
networks, 22–23, 76, 125, 127–128, 130–140, 152–153, 174, 176, 178, 202
Neves, Tancredo, 126
nightingale, 29
Nishida, Toshisada, 127, 201
nonexecutives, 135–137, 177, 203
novel, 10, 11, 43–44, 59
nudity, public, 165

offspring, 6, 7, 12–13, 14, 18, 38, 39, 50, 67, 75–76, 78, 127, 160, 185, 187
openness to experience, 105–106
optimal relationship, 169
options, comparisons of multidimensional, 104

orgasm, female, 44–46, 73, 188
Origin of Species, The (Darwin), 60, 62
oxytocin, 45

parking lot attendants, 149
partible paternity, 74
patent medicine, 31
paternity
  leave, 178–179, 208
  partible, 74
  uncertain, 74, 185
patrilocal societies, 66
Paxson, Christina, 95–96, 194
peacock tail, 29, 49, 50, 122, 180
peahen, 26, 178
penis size, 71
personal contacts, 129–130
persuasion versus force, as male sexual strategy, 58, 72
Pessoa, Fernando, 56, 59
photographs and photography, 15, 44, 68, 165, 188
physical strength, 101, 199
physicians, 100
piece-rate scheme, 114
pill, contraceptive, 53, 99, 195
placebo effect, 55, 119
police state, 161–162, 165–166
politicians, and infidelity, 173
polyandry, 74–75
power, economic, share of, 95
Powers, John E., 32
praying mantis, 8
predator, 28, 49, 73, 77, 157
predictable behavior, 41
preferences, of men and women, 22, 111–112, 117, 121, 128, 133, 136, 138
pregnancy, 42, 161–162, 164, 166, 187
prejudice, 98, 117, 118, 119–120, 136
primatologists, 67, 126, 128
Prince Charming, 152–153
private jets, 37–38
prostitution, 207
psychometric tests, 88, 94, 95, 101–104, 107, 109
psychotherapy, 112
public nudity, 165
puffin, 75

racism, 69
Ramanujan, A. K., 40, 60
reactions between people, testing, 104
"reasonable" agreement, 116
reconciliation, 126
recruitment and recruiters, 106, 115, 130, 176–178
Red and the Black, The (Stendhal), 10, 43
relationship, optimal, 169
relative income, 22
religion, 168
representation of people in companies by height, 93
Ridley, Matt, 73
risk aversion, 115
risk taking, 58, 113–114, 194, 199
rival coalitions, 62
rival males, 13, 19, 29, 52, 58, 65–66
rival theories, 188
rivalry, 13, 65, 66
Roman emperors, 35, 49
Roman empire, 14
Ronsard, Pierre, 141
Ryan, Christopher, 72, 73

sacs, hollow, in dance flies, 29, 31
Sawant, G. P., 141–142
scarce resources, 4–6, 14, 20, 65, 78–79, 82, 84, 118, 145, 153, 161
  eggs, 12–13, 84
  food, 4, 8, 29, 31, 65
scarcity of attention, 145
Schiffer, Claudia, 32–33, 35, 49, 186
scorpion, 7, 13, 73

screever, 141–142
scribe, 85, 141–142, 145
selectivity, 13–14, 16–19, 45, 57, 69, 70, 76, 129, 155–156, 162, 188
self-deception, 55–56, 189
servants, domestic, 164
sex
  appeal, 39, 53
  cells, male, 12, 14
  commercial, 38, 206–207
  consequences of, 6, 7, 162–163
  food and, 2, 4
  homosexual, 6–7
  multiple partners in, 18, 45, 171
  and signaling, 21, 30–31, 39, 43, 49, 53, 165, 190
Sex at Dawn (Ryan and Jetha), 72, 192
sexism, 17, 69
Sexton, Alison, 37, 186
Sexton, Steve, 37, 186
sexual assault, 164
sexual conflict, 7–9, 11–12, 52, 169, 183
sexual partners, 9, 18, 21, 24, 29, 39, 53, 140, 163–164, 168, 170, 186
sexual reproduction, 15, 65, 85
sexual selection, 50, 51–52, 69–70, 108, 129, 140, 151, 191
sexuality, 16, 18, 26, 72–73, 89, 171, 173, 184, 189
"sexy son" hypothesis, 50
Shakespeare, William, 53, 189
share of economic power, 95
shopping, 165
Short, Roger, 66
short-listing, 153, 177–178
shortness, 50, 93, 96–97, 103–104, 118–119
signals and signaling, 10–11, 21, 29
  in competitive sports, 53
  credible, 34–36, 39
  in hygiene rituals, 30
  sexual advertisement 29–31, 38, 49, 52
  and track switch repairers, 147–148
  trap, 123–124, 139, 175–176, 178–179, 181
skin color, 69–70
smile, 25, 27, 46–48, 128, 138, 188
smoking, and lung cancer, 166–168
social codes, 122, 124, 125, 159–161
sociologists, 127, 130, 132, 157
somatic markers hypothesis, 41, 164
song repertoire, 38–39, 187
Sorel, Julien, 43, 59
spatial orientation, 105
spatial reasoning, 68, 103, 109
  and testosterone, 109
spatial skills, 101, 109–110
sperm, 7, 12–15, 20, 29, 58–59, 65–66, 71, 84, 184, 188, 191, 208
  abundance of, 12–13
  competition, 66, 71, 184, 191
  donation, 208
spider (*Agelenopsis aperta*), 8, 13, 58
Stanford, Craig, 67, 191
Stendhal, 10, 43, 59, 187
stereotype threat, 101, 116, 196
Stevenson, Betsey, 156, 193–194, 204
strategy, 19, 29, 50, 52, 55, 74, 175, 189
  sexual, 127
strength, physical, 101, 199
stress, 55, 61–64, 155, 164
strong ties, 127, 129–132
sugar, 24
surgeons, 100, 192
suspicion, 31, 57–59, 163
systematic discrimination, 118

talent, differential, 22
talkativeness, 108–109
tallness, 50, 93, 95–97, 103–104, 118–119, 137, 195
taxi drivers, 100
tea-making, 147
telephone communication, 55, 144, 166

tenth percentile, 147–148
testicle size, 66, 71–72, 191
testing reactions between people, 104
testosterone, 57, 109–110, 195, 198
  and spatial reasoning skills, 109
Thatcher, Margaret, 124
Theory of the Leisure Class, The (Veblen), 35
ties, strong, 127, 129–132
tit-for-tat, 127
Tokyo subway, 158
"torches of freedom" (cigarettes), 99
Toulouse School of Economics, 46, 134
tournaments, 114, 200
Toyota Prius, 36, 37
trade unions, 118
traits
  genetically determined, 107–108, 196
  heritable, 102, 105
Travolta, John, 37
trousers, 99
Trivers, Robert, 55, 184, 189–190, 199
trust
  game, 46–47, 139
  mistrust, 9
  signaling, 21
  and smiling, 46, 47, 49
trusting behavior, 45, 188
tunnel-web spider (*Agelenopsis aperta*), 8, 13, 58
Twitter, 128

uncertainty
  in paternity, 74, 185
  in sexual attraction, 11
unemployment, 91, 194
unjustified exclusion, 95–96, 103
unpredictable behavior, 184
US Bureau of Labor Statistics, 100, 147, 149, 195, 199, 204

vasectomy, 53
vasopressin, 45
Veblen, Thorstein, 35, 186
vegetarians, 170
verbal ability, 101, 196
verbal tests, 68, 103
Veyne, Paul, 35, 186
video and film editors, 148, 149
violence, 14, 26, 83, 89, 90, 160, 171, 187, 193, 205

waiters and waitresses, 101
Wasson, Sam, 171, 206
waste, 53, 145
water strider (*Rheumatobates rileyi*), 7, 13
weak ties, 127, 129–132
Why Sex Matters (Low), 127
Wolfers, Justin, 156, 193–194, 204
women
  negotiating in the workplace, 124, 175
  salaries, 97, 98, 100–101, 116–118, 124, 135–137, 174, 200
  vote, 99
Women Don't Ask (Babcock and Laschever), 116–117
working hours
  flexible, 136
  weekly, 120, 121–122
Working Women Don't Have Wives (Apter), 175
workplace, divided, 168, 174–178
Wrangham, Richard, 64, 77, 190, 191, 192, 205, 226
writers, 11, 141–142, 145, 150

Y chromosome, 90
Yoshino, Kenji, 30, 124
YouTube, 145, 146, 150, 186

zoology, 99